The Day of the Cattleman

By

ERNEST STAPLES OSGOOD

THE UNIVERSITY OF CHICAGO PRESS

CHICAGO AND LONDON

International Standard Book Number: 0-226-63555-4

Library of Congress Catalog Number: 57-7901

THE UNIVERSITY OF CHICAGO PRESS, CHICAGO 60637
The University of Chicago Press, Ltd., London

TO
E. P. O.

PREFACE

To many Americans "the West" is still the land of the unfenced range, the cattleman and the cowboy. Although the last great range herd followed the buffalo and the Indian beyond the horizon years ago, our last frontier lives on in the popular imagination. Its perpetuation may be due to the fact that it was the final chapter in the long story of the westward movement. Perhaps the advertisement it received gave it immortality, for thousands of travelers caught a glimpse of this West from the rear platforms of transcontinental trains. Others, at home, learned of it from Buffalo Bill and his Wild West Show. Finally, in our own day, the western novel and the "western" of the films have kept it alive long after the cowboy had learned the uses of the hay-rake, and had seen the open range give way to enclosed pastures and the great herds of the old days become the bands of cattle around the haystacks of the small rancher.

But the range cattleman has more solid achievements to his credit than the creation of a legend. He was the first to utilize the semi-arid plains. Using the most available natural resource, the native grasses, as a basis, he built up a great and lucrative enterprise, attracted eastern and foreign capital to aid him in the development of a new economic area, stimulated railroad building in order that the product of the ranges might get to an eastern market, and laid the economic foundations of more than one western commonwealth. This is the story that I have endeavored to tell in the following pages.

I am grateful to Dr. Frederic Logan Paxson, whose direction and counsel have been of inestimable value. Thanks are due to my colleagues in the History Department of the University of Minnesota for their interest and encouragement. Finally, to one of Montana's finest pioneers, John Radford, miner, hunter, and rancher, I shall ever be indebted. In his mountain cabin, the Old West lived again.

ERNEST S. OSGOOD

Minneapolis

CONTENTS

CONTENTS

MAPS AND ILLUSTRATIONS

I

THE CATTLEMAN'S FRONTIER
1845-1867

IN 1830, more than two hundred years after the first white man had made a clearing in the forest about him and in so doing had created that most significant of boundaries, the American frontier, the westernmost point in the area of continuous settlement was still less than halfway across the continent. According to the census of that year, the area containing more than two inhabitants to the square mile extended almost as far west as the western border of the young state of Missouri. Here, where the Missouri River coming down from the north bends sharply eastward on its way to the Mississippi, the frontier had paused, and twenty-five years were to elapse before the line of compact settlement advanced much beyond that point. To the rear, north and south, the wings of the frontier line bent far back toward the east and, as the center halted at the bend of the Missouri, the flanks, pivoting on that point, swung slowly westward during the succeeding decades, and new states were formed in the upper Mississippi Valley and in the lower South.

Although the western advance had paused in Missouri, the visitor to the town of Independence, established in 1831 at the apex of this salient, would have found nothing but movement and activity about him. Through its streets and on the river close by, there passed the whole pageantry of the frontier. Here, at the gateway to half a continent, an observer could, as the years went by, mark the emergence of the "Far West," as hunter and trail maker, trapper and trader, home seeker and gold seeker moved out along the western trails into those regions of which the average American was but dimly conscious and about which he knew next to nothing.

The river was a roadway of exploration. Up its lonely reaches had moved the keel boats of Lewis and Clark, a quarter of a century before the founding of the town. Seven years later, the Astorians, whose experiences were to be made familiar to the reading public by the pen of Washington Irving, passed by on their way to the mouth of the Columbia. Then on a day in the spring of 1819, the roving Indian gazed in wonder at a strange monster of smoke and noise moving upstream without any apparent effort on the part of those directing its course. Major Stephen Long and his party on the steamboat *Western Engineer* were on their way to the mouth of the Platte River. From there, in the following spring, they began their journey overland to the heads of the Platte, the Arkansas, and the Red. On his return, Long confirmed the opinions of other travelers that the country beyond the Missouri could never be utilized by white men, but must ever remain the home of the wild tribes who roamed over those frightful and terrifying wastes. For a half-century thereafter the Great American Desert was a fixed idea in the minds of most Americans.

Beyond these "steppes of Tartary," far up in the mountains, the "brigades" of the fur companies and the lonely trapper were busy expanding the great fur trade, which reached its height during the thirties. From the remote north country, where the Missouri and its tributaries head deep in the heart of the Rocky Mountains, they came, their keel boats laden with great bales of peltry for the St. Louis market. Each spring, when the water was high, the inhabitants of Independence turned out to see the steamboat of the American Fur Company, bound for Fort Union, the company's post located at the mouth of the Yellowstone, a thousand miles upstream. As the stories of the "mountain men" circulated around the border settlements and as the journals of explorer and traveler found their way into print, the topography and general character of the mountain regions, hundreds of miles to the west, were known long before the intervening country that began at the outskirts of the Missouri towns was anything more than a name.

This region between the settlements and the mountains, the last area of continental United States to become familiar to the average American, went under the general name of the Indian country. Here was a country, stretching all the way from the Red River to the Canadian boundary, which seemed destined by a kind Providence to provide a permanent home for the Indian. Here he might live undisturbed, freed from the pressure of the westward-moving pioneer, who would never, it was believed at the time, settle in that semi-arid, treeless country where all efforts at agriculture must surely fail. In the western section, on the High Plains and in the mountains, the wild tribes might roam as of old, following the great herds of buffalo upon which their whole tribal existence was based. In the eastern section, close to the Missouri River, room could be provided for the more civilized or the weaker tribes of the eastern United States, who were impeding the advance of the north and south wings of the frontier.

All through the thirties the Federal Government was busy negotiating treaties with these eastern tribes, treaties by which they surrendered their old tribal homes for reservations beyond the western border. When persuasion and solemn promises of undisturbed and perpetual possession failed, force was used, for the western Jacksonian democracy, then in the saddle in Washington, had little patience with humanitarians who demanded that the Indian problem be solved on the basis of abstract justice. Up the Missouri River on steamboats chartered by the government, or along the rough frontier roads of the southwest, the remnants of once powerful tribes moved under military guard to their new homes. Across the border, the new reservations formed an unbroken front from the Mexican boundary at the Red River to the northwestern corner of Missouri. North of Missouri, the tribes of the upper Mississippi were pushed back during the same period, thus clearing the way for the settlement of southern Wisconsin and eastern Iowa.

However permanent and satisfactory this solution of the Indian question might appear to the pioneer farmer and the eastern statesman, the visitor to Independence would soon discover that Indian isolation was the most temporary of expedients. While the treaties were still being negotiated, the wagons of the Santa Fé traders were cutting deeper and deeper the tracks that led out of the streets of Independence, over the sun-baked plains of the Cimarron and the Arkansas, across the Mexican border to the ancient Spanish city where Yankee trade goods could be sold at immense profit. This trade, which flourished during the thirties, quickened the life of the Missouri towns, increased the interest that the border was taking in the Southwest and, incidently, contributed much to the knowledge of the country over which the trail ran.

Before the close of the thirties there were signs of a new movement among the crowds that thronged the streets of the Missouri settlements. In the remote Northwest, beyond the barrier of the Rockies, the American trapper was making contact with the Canadian fur trader in the valleys of the Columbia and its tributaries. Mountain men talked of Oregon, the richest fur country of all, of likely routes thither, and of the necessity for American effort in that region unless it were to become the exclusive domain of the Canadians. In 1832, several parties of fur traders and explorers were outfitting at Independence for the Columbia River. The trail that they took led across the trackless Indian country to the Platte at Grand Island, up that river and its tributary, the Sweetwater, until at last it topped the low divide that separates the waters of the Missouri system from those of the Columbia and the Colorado. Here was South Pass, discovered ten years before by the fur trader, Ashley, a low, grassy divide over which wagons might be drawn with little difficulty. There were no wagon tracks in the year 1832 when Bonneville and Sublette and Wyeth went through, but behind them there was to follow a multitude beneath whose feet rose the dust of the greatest of all frontier roads, the Oregon Trail.

In the history of the westward movement, the missionary has seldom been far behind the explorer and the fur trader, sometimes, indeed, he has led them. In 1834, two Methodist missionaries had established themselves in the valley of the Willamette, a tributary of the Columbia, near Fort Vancouver, where Dr. John McLoughlin ruled benignly over his vassals, white and red, in the interests of the great Hudson's Bay Company. Two years later Dr. Marcus Whitman, sent out by the American Board of Commissioners for Foreign Missions, began his work further up the Columbia in central Oregon. Eastward, over the mountains, in the valley of the Bitter Root, the Jesuits had established themselves by 1840 under the leadership of Father De Smet.

The fertility of the soil was of slight importance to the fur trader. The missionary, however, had a good eye for land, for those Indian converts who could be induced to settle down to farming in the neighborhood of the mission were likely to stay Christianized. In their reports the missionaries were as enthusiastic over the rich land of the Wil-

lamette as they were over the prospect of saving souls. Here
was land that equaled, if it did not surpass, the best that the
prairie region of Illinois could offer. As this news spread,
farmers began to think and talk of Oregon and the way
thither. By 1843 the movement of the homeseeker out over
the Oregon Trail had begun, a movement that in a few
years increased to thousands and built up a new American
commonwealth on the shores of the Pacific. Long lines of
wagons passed through the dusty streets of Independence,
and in the crowd that swarmed around them, the talk was
no longer of fur and Indian trade but of land, of crops, of
climate, and of the fortunes in the fertile soil of Oregon
awaiting those who would brave the long march and all its
attendant dangers.

Two hundred miles upstream, where the Missouri is
joined by the Platte, another group was gathering in the
fall of 1846. In their winter quarters on the western edge
of the new state of Iowa, the Mormons were laying their
plans for the coming spring. They had despaired of finding
a home in the States, for wherever they had settled, their
neighbors had coveted their land, envied their prosperity,
and disapproved of their way of life. Somewhere beyond
the plains and mountains lay the Promised Land. Before the
close of the next year, they had found it in the valley of
the Great Salt Lake.

Then in the next year came the news that was to set the
whole frontier in motion. Eastward along the trail to the
border settlements, across the country to the crowded cities
of the seaboard and on beyond the seas sped the magic
word that was to bring a whole world flocking westward —
gold! The discovery of a few nuggets in a California millrace
was destined to fill the harbor of the Golden Gate with a
forest of masts, to make the Isthmus of Panama a highway
for the nations, and to crowd the Oregon Trail with an

army of adventurers, who would find no rest until the weary miles had been traversed and they stood at last in that fabulous land of gold by the blue waters of the Pacific.

When the emigrant bound for Oregon or California turned his back on the Missouri settlements and struck out along the westward trail, his condition was not unlike that of the traveler sailing out of an eastern seaport on a transatlantic journey. Beyond the narrow wagon track a vast waste stretched away on every side to the far horizon, its swells and hollows as lacking in identity as the crests and troughs of the Atlantic rollers. Herds of buffalo and great bands of antelope, seemingly as multitudinous as the fish of the sea, moved over the face of these great solitudes. It seemed unlikely that man would ever be more than a wayfarer in these wastes. Only the roving Indian, the occasional trapper, and the little garrisons at the trading posts strung out along the trail served to dispel such illusions. The myth of the American Desert, so long a part of the American's stock of ideas about his country, had its origin as much in the impression resulting from such solitary vastness as in any evidence of the sterility of the soil or the rigors of the climate. Men accustomed to the companionship of woods and streams, of green meadows and uplands, of familiar hills and limited horizons, found nothing hospitable in the leagues of brown grass, nothing familiar in the monotony of rolling plain or wind-scarred butte.

Into this great solitude rode the cattleman. From the ranches of Texas and New Mexico he pushed his way northward across the lands of the Indian nations to the railroad that had begun to bridge this waste. The desire for new pastures and markets sent him further and further north, until his herds met and mingled with other herds drifting down out of the northern valleys. It was the range cattleman who broke the spell; who made these great areas his own;

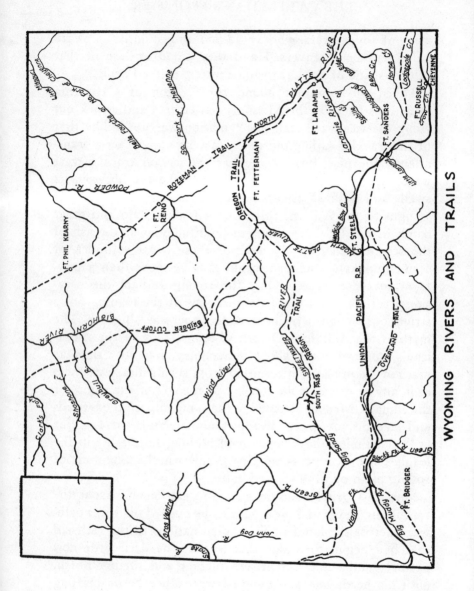

WYOMING RIVERS AND TRAILS

who, in his search for grass, crossed every divide, rode into every coulee, and swam every stream. The solitude of the desert passed, and men began to realize that this, our last frontier, was not a barrier between the river settlements and the mining communities in the mountains but an area valuable in itself, where men might live and prosper.

The cattle business of the High Plains began as a result of the necessities of the emigrants along the Oregon Trail, and the earliest herds were brought together to meet that demand. The westward trek of thousands to Oregon and California in the two decades before the Civil War stirred into new activity the far-western trading posts, which had languished following the boom period of the fur trade. The rather scattered, nebulous population of the fur country began to drift down onto the trail when it became apparent that money could be made out of the western-bound pioneer, who was a ready customer up to the limits of his resources. In these unfamiliar wastes, where nature appeared so strange and formidable to his unaccustomed eyes, he was eager to accept assistance from anyone more experienced than he. By the time he began his journey up the North Platte, his animals were footsore and weak and his stock of food was running low. It was a strong and well-equipped outfit indeed that was not anxious to bargain for such aid and comfort as those along the trail were able to furnish.

Nor were the traders who were finding favorable locations along the trail loath to gain all they could from these necessitous travelers. Flour, coffee, bacon, powder, and shot were always in demand. Sometimes the emigrant lacked these essentials because of ill-advised provisioning at the outset, sometimes he was the victim of wandering bands of Indians who held up trains and exacted tribute. Flour, brought down by packhorse from the Oregon settlements, sold for one hundred dollars a hundredweight on the trail.[1] As early as 1845 Fort Bridger had become one of the chief entrepôts of this trade. Hither the mountaineers had resorted for years to trade their season's supply of hides and Indian articles for flour, pork, spirits, powder, lead, blankets, butcher knives, hats, ready-made clothes, coffee, and sugar.[2] Such posts merely had to enlarge their stocks in these articles to meet the emigrant's demands.

But the traders soon found ways of making money other than by selling these standard supplies of the posts. Three new economic activities sprang up along the trail, each of them the result of utilizing the local natural advantages and resources and each of them a part of the business of transportation rather than supply. These were the operation of bridges and ferries, the furnishing of forage, and the exchanging of fresh for worn-down work cattle.

It was not long after the western migration had begun before bridges or ferries were established at the more difficult stream crossings. At strategically located points on the North Platte, the Sweetwater, and the Green rivers, ferrymen were prepared to take the emigrant and his team across for a toll.[3] These ferries became natural trading points, and here road ranches, often the property of the ferryman, sprang up.

With every year of travel over the emigrant road, it became more and more difficult to find sufficient grazing

---

[1] Joel Palmer, "Journal of Travels over the Rocky Mountains, 1845-1846," *Early Western Travels* (Cleveland, 1906), edited by R. G. Thwaites, XXX, 86 *et seq.*

[2] *Ibid.*, 74-75.

[3] The toll bridge over the North Platte, twenty miles west of Fort Laramie, which cost $5,000 to build, took in $40,000 in tolls during the single season of 1853. A five-gallon keg of whisky was sufficient to pay a toll charge of $125.00 on a train of nineteen wagons crossing the Platte at this point. "Autobiography of William K. Sloan," *Annals of Wyoming* (Cheyenne) IV, 246, July, 1926.

ground for the animals. As a result, there developed a market for hay. Temporary posts, consisting of a tent and a corral set up along the trail to catch the season's trade, were soon converted into more substantial ranches. Their owners began to put up the wild hay that grew along the streams and were prepared to supply forage to the motive power of the emigrant trains at thirty-five cents to a dollar and a half a hundredweight. A small garden patch on the side might prove profitable, when potatoes brought five cents apiece during the emigrant season. Such establishments usually consisted of an adobe house, often a dwelling and store combined, a few stock corrals made out of the cottonwoods that bordered every stream, and a haystack.[4] These road ranches, the product of the emigrant trade, were the first ranches of the northern ranges.

The need of the travelers for fresh work stock and the profits to be made out of such a trade induced many of the traders to go into the cattle business. One fat and well-conditioned work steer might be exchanged for two worn-down and footsore ones. Dairy cattle, driven along with the trains, appeared less valuable on the Sweetwater than they did in Missouri, and many a family cow, unused to the hardship that such a journey imposed, was destined never to reach the green valleys of the Willamette but was traded off for ten dollars or a little flour.[5]

The early herds of the northern ranges were the product

[4] Diary kept by Silas L. Hopper, "Nebraska City to California, April-August, 1863," *Annals of Wyoming*, III, 117, Oct., 1925. Gen. Sherman on his trip west in 1866 wrote back to Rawlins that "these ranches consist usually of a store, a house, a corral, and a big pile of hay for sale . . . you are never out of sight of a train or ranch." Sherman to Rawlins, Aug. 21, 1866, *House Ex. Doc.* No. 23, 39 Cong., Sess. 2, p. 5.

[5] Sometimes this loose stock amounted to a considerable band. The good price for beef at the California mines induced some herdsmen to essay the long drive with a beef herd. Greeley notes such a herd from southwestern Missouri. Horace Greeley, *Overland Journey* (New York, 1860), 72.

of such trade. Captain Richard Grant, trading along the road from Fort Hall, had a herd of six hundred in 1856.[6] Horace Greeley, on his way to Salt Lake three years later, found this business thriving along Black's Fork and Ham's Fork of the Green River. Here he found "several old mountaineers, who have large herds of cattle which they are rapidly increasing by a lucrative traffic with the emigrants, who are compelled to exchange their tired, gaunt oxen and steers for fresh ones on almost any terms. R. D., whose tent we passed last evening, is said to have six or eight hundred head; and, knowing the country perfectly, finds no difficulty in keeping them through summer and winter by frequently shifting them from place to place over a circuit of thirty or forty miles. J. R., who has been here some twenty odd years, began with little or nothing and had quietly accumulated some fifty horses, three or four hundred head of neat cattle, three squaws, and any number of half-breed children. He is said to be worth $75,000." [7] These were Wyoming's first cattlemen.

As the forage along the trail became scarce from constant cropping, the more enterprising herdsmen drove their cattle north into the sheltered valleys of the upper Missouri in what later became western Montana, their wintering places being the Beaverhead, the Stinking Water (later the Ruby), and the Deer Lodge valleys. The value of this region as a stock-raising country had been demonstrated by the Jesuit fathers at the St. Ignatius Mission, located on the Clark's Fork of the Columbia. Here under their tutelage, the Flatheads had settled down to a more or less civilized existence and by 1858 had developed so far in the arts of farming and animal husbandry that they were sowing three

---

[6] Granville Stuart, *Forty Years on the Frontier* (Cleveland, 1925), II, 97.
[7] Greeley, 195. This entry was made while Greeley was at Fort Bridger. The J. R. referred to may have been J. B. — Jim Bridger.

hundred acres to wheat and were herding on the adjacent
hillsides and in the neighboring valleys over a thousand
head of fine stock.[8]

Had it not been for the Mormon war of 1857-1858, the
Jesuits and their Indian converts might have remained
undisturbed for another decade. When, however, the elders
of the Mormon church issued an edict in February, 1857,
ordering the Gentiles within the Mormon territory to leave
forthwith, the isolation of the mountain regions north of
the trail was destroyed. During the years previous to 1857,
many enterprising merchants from the Missouri river towns
had brought out loads of goods and had set up in business
in the Mormon settlements. This trade had proved enormously profitable and considerable sums had been invested
in the business. The order to evacuate Mormon territory
left these merchants with no alternative than that of immediately disposing of their stocks as best they could. Many
of them traded off their remaining merchandise for the
cattle of the Mormons at ruinous figures and hurried out of
the territory before their enterprising customers could recover the purchase price by stampeding the herd. Some
headed for California where the mining communities offered
a safe market. Others drove northward to the posts along
the trail.[9] Here traffic had stopped when the rumors of

---

[8] Report of Lieutenant B. F. Ficklin to Major F. J. Porter, April 15, 1858,
in Annual Report of the Secretary of War, 1859, *House Ex. Doc.* No. 2,
35 Cong., Sess. 2, Vol. II, pt. 2, p. 70. Major John Owen had in 1850 purchased the buildings of St. Mary's Mission on the Bitter Root River from the
Jesuits. This mission had been established nine years before by Father De
Smet. Owen established a trading post here that he called Fort Owen. When
the early cattlemen entered the valley from the south, they found Owen cultivating a considerable plot of ground and pasturing stock that he had
bought of the Catholic fathers. Paul C. Phillips, *The Journals and Letters of
John Owen* (New York, 1927).

[9] Sloan, "Autobiography," *op. cit.*, 260-263. Sloan was engaged in this
trade with the Mormons. He had a store at Provo and was driven out along

burned freight trains and massacred emigrants sped eastward. The traders, seeing their custom diminish and fearing
the ravages of the Saints and their Indian allies, sought
refuge in the mountains until the storm blew over. Into the
valleys of western Montana straggled the herds of the
traders and of those who had been expelled from Utah.

Neither the protection afforded by the army of General
Albert Sidney Johnson sent out to quell the rebellion, nor
the market for beef, which the presence of this force created,
was sufficient to tempt the traders to come down out of the
northern valleys. In December, 1857, a small detail from
Johnson's forces was sent north to contract for beef with
these fugitive cattlemen. The report of the commander of
this beef-buying expedition gives a good picture of the situation in the upper Missouri country, the cradle of the stock-
growing industry of Montana.

After experiencing great difficulty in crossing the snow-
choked divide that separated the headwaters of the Missouri from those of the Snake, the party got down into the
upper Missouri country.

After getting on the head waters of the Missouri, the snow entirely
disappeared. On the fourth, our rations were exhausted, but I was
not uneasy, as I expected to arrive soon at the Beaver Head, a point
on the Jefferson Fork of the Missouri, fifty miles above the Three
Forks of the Missouri, and one hundred east of the Mormon settlement on Salmon River, a popular wintering ground of the mountaineers, on account of their stock.

To my surprise, on arriving at Beaver Head, I found all the evidences of the mountaineers having left recently, and hastily, and
taken the trail in the direction of Flathead valley. . . .

On the 10th, overtook the camp of Mr. Herriford, where I obtained a supply of beef, and learned from him that about December

with the other Gentiles in the Territory in 1857. He estimated the total
Gentile population at about three hundred in Salt Lake and not more than
fifty in the rest of the Territory.

first they had heard of the burning of the supply trains by the
Mormons, and of threats uttered by the Mormons at Salmon river
fork, against the mountaineers at Beaver Head. Fearing for the
safety of their stock, they had started for the Flathead valley, as a
more distant and secure point.

At the Deer's Lodge, overtook another party of mountaineers,
with whom I made a contract for the delivery of three hundred head
of beef-cattle, by April 16th, at ten dollars per hundred [weight],
also to bring down about one hundred head of horses. Afterwards
proceeded down the Flathead valley, where I could have a contract
for two hundred head of cattle, but their fear of the Mormons was
so great that no price would induce them to undertake to deliver them
here. Several were making preparations to move their stock to Fort
Walla-Walla this spring, in order to be beyond the reach of the
Mormons. . . .

I spent several days at St. Ignatius mission (situated on one of the
branches of Clark's Fork of the Columbia, on forty-seventh parallel)
established by the Catholics, for the benefit of the Flatheads, Pend
d'Oreilles, and Hootenais [sic].

. . . . Under the direction of the priests they are improving rapidly
in agriculture. This year they will sow about three hundred bushels
of wheat; they raise large quantities of vegetables, especially potatoes,
cabbage, and beets.

Their horses are superior to all other Indian horses, in size and
power of endurance. The tribe, about sixty lodges, owns about one
thousand head of cattle.

As it was impossible to buy stock in Flathead valley, on conditions
contemplated in my instructions, on March 3rd I started for Deer
Lodge, expecting to start immediately on my arrival with what stock
I had contracted for at that place.

The contractors refused to deliver their beef at this place [Fort
Scott, Utah] but offered to deliver it there [Dear Lodge Valley]
as they were afraid of being robbed by the Mormons on the road.

Buying a few animals, to replace those lost, started on March 12th
to return, . . .

The new grass was beginning to grow finely before I started on

Jefferson fork; contrary to my expectation and information I had received from the oldest mountaineers, found snow in the mountains, between Missouri and Snake rivers, from three to six feet deep for a distance of twelve miles. . . . . [10]

The Mormon danger was, however, only temporary, and in the following year the trade along the trail was as brisk as ever. The sojourn of the traders in the mountain valleys had given them much information of the grazing resources of the upper Missouri country and had established a practical route from the trail to that region. Later, when gold was discovered in western Montana, the trail over which the traders fled with their herds became the chief connection between the mining towns of Montana and the great central route of transcontinental travel.

In addition to these herds of the traders, which had had their origin in the trade along the emigrant trail, there were the train-cattle or "bull-teams" of the freighting companies, which supplied the army on the plains, brought out the Indian annuity goods, and furnished the mining camps in the mountains with the necessities of life and equipment for the mines.[11] These trains of thirty or more wagons to a unit, each wagon with its six yoke of oxen, creaked their way across the plains in an almost endless procession. Thousands of head of these work animals were wintered by their owners in favorable spots along the trail. In the winter of 1857-1858, the firm of Russell, Majors, and Wadell wintered fifteen thousand head on a range that extended southward from the trail for a distance of over two hundred miles.[12]

[10] Ficklin Report, op. cit., 69-70. See M. L. Wilson, "Early Montana Agriculture," Proceedings of the Mississippi Valley Historical Association, 1918, IX, 429-440; also Conrad Kohrs, "A Veteran's Experience in the Western Cattle Trade," Breeder's Gazette (Chicago), Dec. 18, 1912, pp. 1328-29.

[11] Frederic L. Paxson, History of the American Frontier, 462.

[12] Annual Report of the Commissioner of Agriculture, 1870, pp. 303-309.

This range was far enough east so that the Mormon danger was not felt.

The experience of the early cattlemen along the trail and in the mountains of western Montana had demonstrated the practicability of wintering stock on the northern ranges a full decade before the Texas longhorn put in his appearance. Any further expansion in this pioneer industry beyond the point already described had to wait on the development of new local markets.

The discovery of gold in the Rocky Mountains, coincident with the Mormon outbreak and the scattering of the herds into the mountain valleys, created just such a market. In the autumn of 1858 gold was discovered some two hundred miles south of the Oregon Trail on the upper waters of the South Platte. By the next spring, the plains were alive with the Pike's Peak gold rush. The old trail was crowded, and to the south other thousands of gold seekers were making new trails across the unfamiliar brown wastes to where rise the eastern escarpments of the Rockies. The oxen used for this new trek were turned out to graze on the plains at the foot of the mountains, while their owners hurried on up the canyons to the diggings. For the more thrifty, ranches were established where cattle could be boarded for a dollar and a half a month.[13]

Here was a local market, which must be supplied, and which, in the fever of the gold rush, was not inclined to haggle over the price. The winter of 1858-1859 saw twenty-five thousand people at the Colorado mines or on the road, and beef of any kind or quality was at a premium. "From that time to the present," commented the *Rocky Mountain News* in retrospect twelve years later, "the Denver market has been supplied exclusively the year around with beef from

[13] Greeley, *op. cit.*, 115.

the neighboring plains." [14] Train cattle and the stock of the gold seekers were used to start the ranches that began to grow up along the South Platte. In 1861 Iliff, destined to become the first "cattle king" of the northern ranges, was supplying the Colorado mining towns with beef from a herd that ranged up and down the South Platte for a distance of seventy-five miles or more.[15]

In another region the stimulus of this new and insistent market was being felt. Close to the southern borders of Colorado Territory, small communities of Mexicans had settled along the upper Rio Grande and its tributaries. Here they developed a system of stock growing perfectly adapted to their physical environment, a system that the cattle growers of the High Plains were never able to duplicate because of the inadaptability of eastern-made land laws. "They hold their lands," wrote one observer, "without title and in accordance with their own customs. The land along the streams, being the only land that can be cultivated, each man holds so many varas or yards front on the stream and extending back at right angles with the stream to the bluff or as far as water can be carried by ditches for irrigation. The rest of the land is open to all as pasture and worthless for any other purpose. By this system of survey, each man has an equal use of water and bottom land, whether he cultivates three varas or one hundred, and all would be willing to pay for the land cultivated if they could take it in the *shape* they now hold it. The survey and sale of this land in regular sections would probably drive out the present population, while it might fail to bring in an equally industrious one." [16]

14 The *Rocky Mountain News* (Denver), quoted in the *National Live Stock Journal* (Chicago), I, 71, Nov., 1870.

15 Dr. Henry Latham, *Trans-Missouri Stock Raising; The Pasture Lands of North America* (Omaha, 1871), 41.

16 Report of the Surveyor-General of Colorado, Utah, Nevada, and Idaho

Cattle from these ranches found a ready sale in the Colorado towns, and thus the first connection between the southern stock-growing areas and the northern ranges was established, a connection that was to grow in magnitude until it constituted one of the most distinctive features of the "cow country."

The "busted" gold seekers of the Pike's Peak rush had scattered by 1862. Some had limped back to the border settlements to form an outer crust of plains-wise folk along the Kansas and Nebraska frontier; some drifted into the freighting business on the trails or took to ranching along the Platte or on the upper reaches of the Arkansas; some followed the rumor of gold to the north and became denizens of the roaring mining camps of the Clearwater and Salmon rivers. To the east, across the Bitter Root Range, some of the herdsmen who had fled from the Mormon danger were finding pay dirt in the Deer Lodge Valley.[17] News of these strikes filtered into the camps to the west and south. In 1862 a wave of prospectors rolled through the western passes, and by 1865 Bannock, Virginia City, and Helena were all on the map.

The solitary prospector might live off the country. As he worked from one mountain gulch to another, the bands of elk, blacktail, and mountain sheep furnished him with his chief food staple. Groups of miners, for whom the season had not been successful, often wintered in some likely hunting country and not uncommonly got through the winter on a bill of fare of "meat straight." Gathered in the mining camps by the thousand, however, they must be fed, and all the necessities of life, save what the country could supply, must be freighted in.

in the *Annual Report of the Commissioner of the General Land Office*, 1864, p. 80.

[17] Stuart, I, 132-156.

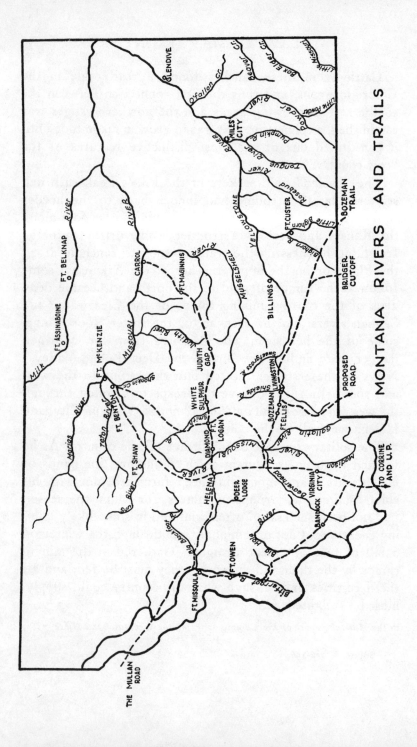

MONTANA RIVERS AND TRAILS

Here was a market for the Montana stock grower, who soon found that taking gold dust from the miners in exchange for beef was almost as profitable and far more certain than getting it from the placers. Even a poor worndown ox might bring one hundred dollars in gold when its owner auctioned it off to the Sunday crowd of miners in the street of Virginia City where beef sold on the butcher's block at twenty-five cents a pound.[18]

Such prices as these and the free pastures in the mountain valleys induced many of the new arrivals to engage in stock raising. A demand was thus created for stock cattle, which was felt in Oregon, California, on the Platte, in the border settlements of Kansas and Missouri, and even in Texas. As early as 1866, Nelson Story came up over the Bozeman Trail to the Gallatin Valley with a herd of six hundred Texas longhorns that he had picked up in Dallas.[19]

The number of cattle in the vicinity of the mines increased rapidly. By 1868, five years after the settlement of Virginia City, the assessors of the nine counties of Montana listed 10,714 oxen and 18,801 cows and calves. Four years later, although the number of oxen had fallen off, because of the practice of using mules and horses for freighting, the number of stock cattle had risen to over 75,000.[20] In Deer Lodge County, the center of the new industry, cattle had become so numerous that the need for regulating the winter range was felt. The fact that the Federal Government possessed the sole power to legislate for the public domain did not prevent the Montana territorial legislature in 1866 from passing a law giving the county commissioners of Deer Lodge County power to define what should be summer grass land in the county and prohibiting stock owners from pasturing their

---

[18] Kohrs, 1328.

[19] A. L. Stone, *Following the Old Trails* (Missoula, 1913), 212.

[20] Annual reports of the auditor and treasurer of Montana Territory, Helena, 1860-1872.

stock on winter grass land, unless they owned the same.[21] Although this law was repealed the next year, it is significant, for it illustrates how soon after the establishment of the stock-growing industry in a given area, the problem arose of conserving the free grazing of the public domain. As we shall see, neither the stockman nor the government was able to solve the problem.

The settlement of a large mining population in the mountains, the resulting increase in traffic across the plains, and the building of the Union Pacific, all occurring between 1860 and 1870, rudely disturbed the Indian isolation of the preceding decade. The Indian hostilities that ensued forced the Government to give more attention to the military problem of the plains, and resulted in the establishment of forts to protect the new communities and the various lines of overland communication. These new army posts created additional local markets where good prices were paid for beef. In 1871, the newly established post at Cheyenne, Fort D. A. Russell, was paying a contract price of eight dollars and thirty-five cents a hundredweight to the cattlemen along the Platte.[22] Much of the trade for the early ranchers of Wyoming centered around these forts, where quantities of hay for the cavalry mounts and beef for the men, two commodities that the locality was prepared to supply, were needed.

In 1867 the rails of the Union Pacific penetrated Wyoming. The work gangs who laid the rails and the horde of hangers-on who constituted those ephemeral towns at the rail head must be fed. Buffalo, brought down by such hired

[21] *Laws of Montana Territory,* 1866, Sess. 2, p. 35. This law, which was an invasion of the power of the Federal Government over the public domain, was repealed at the next session. *Laws of Montana Territory,* 1866, Sess. 3, p. 83.

[22] Letter of T. H. Durbin in *Letters from Old Friends and Members of the Wyoming Stock Growers' Association* (Cheyenne, 1923), 45.

men of the railroad as Buffalo Bill, helped to meet this demand, but the cattle of the Wyoming ranchman found as ready a market along this first transcontinental railroad as they had found along the old emigrant trail.

Thus, by the close of the sixties, there existed in the northern section of the High Plains and in the adjacent mountain valleys, herds of considerable size, recruited from the stock of the emigrant and gold seeker, from the work animals of the freighting companies, from the Mormon herds, and from the herds of Oregon and California. Their owners were making good profits in supplying the local market of mining camp, section crew, and military post. The possibility of expanding their herds so as to utilize to the full the enormous pastoral resources on every hand depended upon a supply of cheap cattle that could be used for stocking the empty ranges and upon a connection with the eastern market.

The inhabitants of the brash little towns on the Union Pacific were conscious that they were living along one of the great highways of the world's commerce. They speculated on the wealth of the rich cargoes from the Orient, borne eastward by long lines of freight cars. Local newspapers noted in their columns the passing of especially valuable trainloads of tea and silk from China or ore from the mines, and commented upon the fact that fortunes were rolling by their very doors every day. Out on the Laramie Plains and along the tributaries of the Platte a less romantic way freight was developing, far more essential to the well-being of these communities and of the railroad that served them. The passing of the first stock train bound for the Chicago market meant that the utilization of these northern ranges had begun in earnest.

II

THE TEXAS INVASION

ALTHOUGH the early cattlemen along the trails and adjacent to the mining communities had demonstrated that the High Plains were grazing grounds of enormous potentialities, the notion of the Great American Desert, which every school geography had fixed in the minds of two generations of Americans, still persisted.[1] Writing in 1868, the Commissioner of the General Land Office stated the general opinion as follows:

"Although the barrier [the High Plains] is not an unsurmountable one, the fact nevertheless remains, that this belt of country is an obstacle to the progress of the nation's growth — an impediment to the prosperity of the new communities west of it, in not yielding that sustenance required for increasing population."[2]

While the Commissioner was reasserting the dogma of the Great American Desert, the process of utilizing this area got under way. On the western edge and close to the mountains, the cattlemen, whose market had been the mining camps, were prepared to move down the valleys and out into the open country. In southern Wyoming, the herds that had supplied the trail and the army posts were ready to expand into the country north and south of the overland routes. Along the eastern border, the frontier farmer in Nebraska was working up the valleys of the Platte and the Missouri,

[1] R. C. Morris, "The Notion of a Great American Desert East of the Rockies," *Mississippi Valley Historical Review*, XIII, 190-200, Sept., 1926.

[2] *Annual Report of the Commissioner of the General Land Office*, 1868, p. 138.

north and west of Omaha. In Kansas, settlement had expanded as far west as Junction City on the Kansas River by 1860.[3] During the decade that followed, the whole eastern third of the state had been settled, so that the census of 1870 showed the line of settlement of more than two to the square mile close up to the Big Blue River to the north, thence southwest to Abilene, thence south to where the Arkansas bends southward at Wichita.

But the pioneer stockman of Colorado, Wyoming, and Montana would have had to be satisfied with a restricted local market, and the frontier farmer in Kansas and Nebraska would have had to put up with no market at all, if the western railroad, that one essential upon which the whole future utilization of this new frontier rested, had not been in the making. To those promoting railroad construction, the utilization of the region west of the one hundredth meridian by either of these two groups seemed remote. As it had been with the emigrant bound for Oregon, so now with the railroad builders, the High Plains were merely areas to be traversed on the way to the mining camps and to the Pacific. Although a decade later, Wyoming steers and Nebraska hogs and wheat meant more to the officers of the Union Pacific than chests of tea and bolts of silk from the China trade, or bars of gold or silver from the mines, in 1867 they might well disregard the possibilities of picking up any considerable body of eastern-bound way freight between the mountains and the Missouri. To the south, the Kansas Pacific, which had built to a point some fifty miles west of Abilene on the Kansas River in 1867, might figure on the business of the farmers along the eastern end and on the Denver trade, whenever the road reached that mining center.

One more element in combination with the mountain

[3] Greeley, *Overland Journal*, 73.

herds, the frontier farmer, and the western railroad determined the course that the economic development of the High
Plains was to take. Across the land of the Five Nations,
herds of Texas longhorns were swinging northward in 1867.
These would fill the eastern-bound freight cars and develop
a way traffic of enormous value to the railroads. They
would occupy western Kansas and Nebraska until the westward-moving farmer fenced them out, and finally they would
serve to stock the northern ranges.

In all the varied history of our frontier, no single activity
attracted more attention from contemporaries nor called
forth a greater flood of reminiscence, story, and song after
it had passed than did the Texas drive. To those who took
part, accustomed as they became to all the possible incidents
of the drive, familiar as they were with the solitudes over
which they passed, each drive was a new adventure and its
successful completion always brought to the most experienced something of the thrill of achievement. In after years,
the drive of the Texas men became little short of an American saga. To all those who saw that long line of Texas cattle
come up over a rise in the prairie, nostrils wide for the smell
of water, dust-caked and gaunt, so ready to break from the
nervous control of the riders strung out along the flanks of
the herd, there came the feeling that in this spectacle there
was something elemental, something resistless, something
perfectly in keeping with the unconquered land about them.[4]

The Texas longhorn was of no undistinguished ancestry
in spite of his appearance. "They are, indeed," commented
a Scotch observer, accustomed to sleek Polled Angus and
Aberdeen, "nothing else than Spanish cattle, direct descend-
ants of those unseemly, rough, lanky, long-horned animals,

---

[4] In *The Trail Drivers of Texas* (2nd ed., Nashville, 1925), J. M. Hunter
has brought together the reminiscences of hundreds of those who went up
the trail.

reared for so long and in such large herds by the Moors
on the plains of Andalusia." [5] They were of light carcass
with long legs, sloping ribs, thin loins and rumps, and a
disproportionately large belly. In color they were nonde-
script, yellow, red, dun, and black, with often an iron-grey
stripe along the back. Their meat was coarse and stringy,
"teasingly tough." They were almost as wild as the buffalo
that they supplanted on the plains, for behind them were
generations of untamed ancestors. To drive a herd of such
beasts, to work them over, to brand and ship them to market meant a business that would not be without adventure
and danger.

The States had known of the Texas longhorn before the
Civil War. Now and then the herd of some enterprising
Texan would arrive in the river towns, St. Louis, St. Joseph,
or Kansas City. In 1857, a herd got as far north as Quincy,
Illinois, where it was sold for feeders to the local farmers.[6]
The Morgan line of steamships, which plied between Gal-
veston and New Orleans and Mobile, loaded on as many
of these cattle as could be crowded between decks, when
other and more profitable freight was lacking. The company
itself paid for these and sold them in New Orleans and
Mobile, for the freight charged to anyone outside the
company was so prohibitive that it precluded all private
shipment.[7] The plantations of Cuba could use beef if it
were cheap enough, and now and then a shipload of longhorns was sold in Cuban ports.

During the war, a few herds were driven across the
Mississippi, and the Confederate soldier had the opportunity to test his fortitude on Texas beef. When the Federal
forces got control of the line of the Mississippi, even this

[5] James MacDonald, *Food from the Far West* (London, 1878), 30.
[6] J. G. McCoy, *Historic Sketches of the Cattle Trade of the West and
Southwest* (Kansas City, 1874), 88.
[7] *Ibid.*, 19.

limited market was closed. Cattle were almost worthless. In the central portions of the state, they roamed at will, their owners scarcely knowing where their property was nor particularly interested in its increase. Cotton was the great crop of Texas, but the condition of the cotton grower was no better than that of the cattle raiser, for with the war his market also had disappeared. Enormously rich in potential resources, Texas must wait for better times.

With the close of the war the possibility of finding a market for the Texas steer did not appear quite so hopeless. In the first place, the price of beef was high and the supply was not by any means equal to the demand. There were more people in the country and fewer cattle. In the decade, 1860-1870, the population of the United States had increased 22 per cent but there had been a decrease in the total number of cattle (milch cows, work oxen, and other cattle) of 7 per cent.[8] Again, the building of the Kansas Pacific westward created a road to a market anxious for beef of any quality. Finally, the northern plains were waiting to be stocked.

In 1866 the Texas cattle owners, who were usually small cotton and corn farmers as well as stock growers, were stirring themselves, claiming possession of their long neglected stock, which roamed at large beyond the fenced-in fields. "Cow-hunts," the primitive forerunners of the roundup, were staged all over the eastern and central parts of the state. A description of one of these early roundups gives an

---

[8] *Ninth Census*, III, 82, 87. The census figures in 1860 and 1870 for cattle are as follows:

	1860	1870	% Dec. or Inc.
Milch Cows	8,585,735	8,935,332	4% Inc.
Work Oxen	2,254,911	1,319,271	41% Dec.
Other Cattle	14,779,373	13,566,005	8% Dec.
	25,620,019	23,820,608	7% Dec.

The decrease was greatest in the number of work oxen, due partly to the war and partly to the abandoning of the work ox for transportation purposes.

interesting picture of the condition of the livestock industry in Texas at this time:

We didn't call it roundup in those days. We called it cow-hunts and every man on this cow-hunt was a cattle owner just home from the war and went out to see what they had left and to brand up. . . . I was the only boy on this cow-hunt . . . and was looking out for cattle that belonged to my father before the war. We had no wagon [chuck wagon]. Every man carried his grub in a wallet on behind his saddle and his bed under his saddle. I was put on herd and kept on herd when we had one and I don't think there was ever a day on this hunt when we didn't have a herd, and I carried a lot of extra wallets on behind my saddle and a string of tin cups on a hobble around my pony's neck. A wallet is a sack with both ends sewed up with the mouth of the sack in the middle. I just mention this for fear some of the cow men don't know what a wallet is. Whenever the boss herder couldn't hear those cups jingling, he would come around and wake me up. We would corral the cattle every night at some one of the owners' homes and stand guard around the corral. I didn't stand any guard but I carried brush and corn-stalks and anything I could get to make a light for those who were on guard to play poker by. They played for unbranded cattle, yearlings at fifty cents a head and the top price for any class was $5 a head, so if anyone run out of cattle and had a little money, he could get back into the game. For $10 say, he could get a stack of yearlings. My compensation for light was twenty-five cents per night or as long as the game lasted. Every few days they would divide up and brand and each man take his cattle home, as we now call it — throw back.

This cow-hunt continued all summer and when I wasn't going to school, I was picking cotton or cow-hunting.[9]

[9] From a letter by Lee Moore, Texas trail driver, foreman of a Wyoming outfit in the 80's and present secretary of the Wyoming Board of Live Stock Commissioners, found in *Letters from Old Friends and Members*, 33-34.

From the descriptions of Texas of this period, it would appear that the method of pasturing stock was not so very different from what one would find in early Illinois or in Missouri, except that the whole enterprise was on a larger scale. The cattle roamed at will in a country that was by no means devoted solely to cattle raising, as was later the case on the northern

It was in this year, 1866, that there occurred the first considerable movement of Texas cattle northward in an attempt to reach the region of good prices. It is difficult to estimate the size of this drive, because it did not arrive at any railroad shipping point or points, but was scattered chiefly into the feeding areas of Kansas, Missouri, and Iowa. McCoy, the chief authority we have for this early drive, states that it was rumored that no less than 260,000 head set out for the north. About all we know is that by the summer of 1866, herds of considerable size were crossing the Red, in some cases in charge of the owners, in others, in the hands of drovers who "put up" a herd, taking the cattle on credit and giving a list of brands and amounts due to the owners.[10]

The route taken by this drive began at the crossing of the Red at Colbert's Ferry, traversed the southeast corner of the Indian country to Fort Smith, Arkansas, thence northward through the Ozarks and across southern Missouri to the line of the Missouri Pacific Railroad at Sedalia, Missouri. A branch of the trail led westward into eastern Kansas.[11] The drovers found difficulties a-plenty. The Indian Territory was in such turmoil as a result of the division of the tribes during the Civil War that there was a good chance of the cattleman losing a part or all of his herd at the

range and in western Texas. An Iowan down to buy cattle in 1866 described the country along the Brazos, one of the chief stock-raising areas, as having fine farms where cotton and corn were raised and where the "prairie was literally covered with tens of thousands of cattle, horses, and mules." John Duffield, "Driving Cattle from Texas to Iowa, 1866," *Annals of Iowa*, XIV, 249, April, 1924.

[10] McCoy, 20-29, 78-79; G. W. Saunders, "Origin of the Northern Trail," in *The Trail Drivers of Texas*, 20-25.

[11] "Cattle Trails in Live Stock Market Development," in *The Monthly Letter to Animal Husbandmen*, published by Armour's Livestock Bureau (Chicago), Vol. VII, No. 1, April, 1926. The map accompanying this report gives all of the Texas cattle trails.

hands of the Indians or outlaws who roamed uncontrolled over the country.[12]

Much of the country traversed by this first drive was rough and timbered, particularly the Ozarks, and here the drovers learned that Texas cattle must be kept away from wooded country, for in timber they become unmanageable. The farmers of southern Missouri and Kansas were out in force to repel this horde of wild beasts, which stampeded across their farms, and which threatened their own stock with the Spanish fever. Thus for various reasons the drive was held up, and by midsummer the country in the vicinity of Baxter's Springs in the southeastern corner of Kansas was alive with cattle. Some got to the railroad and were shipped to Chicago, others were sold to the farmers along the route northward. Some got as far north as Iowa, where the farmers of that section bought them up as feeders, paying as high as $35 a head, an almost unbelievable price to a Texas cattle owner in 1866.

In 1867 the Kansas Pacific had reached a point some two hundred miles west of Kansas City. It was now possible to reach a railroad west of the line of the drive of 1866, thus obviating some of the difficulties that had rendered the drive of that year more or less of a failure. Crossing the Red some fifty miles further to the west, the herds of 1867 followed the most famous of the cattle trails, the Chisholm Trail, and came out on the Kansas Pacific at Abilene. The figures for this drive have been placed anywhere from 36,000 to 75,000 head.[13] At any rate, whatever the number may have been, the northward movement was definitely

[12] *Annual Report of the Commissioner of Indian Affairs*, 1865, pp. 32-42; Annie Heloise Abel, *The American Indian under Reconstruction* (Cleveland, 1925), Chap. III, "Cattle Driving in the Indian Territory," 73-97.

[13] The following figures for the drives from 1866 to 1885 are given in the *Second Annual Report of the Bureau of Animal Industry*, 1885, p. 300:

established in 1867, and from then on, each spring found the Texas cattlemen "putting up" larger and larger herds for the northern market.

In the late sixties and early seventies the business was haphazard enough and reflected the general opinion of Texas cattle owners as to the value of their property. Ranchmen gathered up the most likely of their beef stock and turned them over to the drovers. Often drovers did not put the owners to the trouble of a roundup, but went out on the range and collected such cattle as appeared saleable without reference to ownership. "When collected," to quote one observer, "he [the drover] would examine and make a memorandum of the brands borne by the cattle. Each year there is held a 'stock meeting' at which all the stock raisers, cattle drovers and traders come together and those who have driven cattle out submit a memorandum of the brands on the cattle taken out by them and settle up with the owners of the brands. If an outsider came and wished to buy up a herd, he would hunt around for someone who would sell him two or three thousand head. . . . The terms agreed upon, the seller would go out upon the range, drive up the required number of cattle, without reference to who owned them, and turn them over to the outsider who would drive them north. At the annual meeting, the seller would report the number of cattle and the several brands and make settlement with the owners." [14]

1866 — 260,000	1873 — 404,000	1880 — 394,784
1867 — 35,000	1874 — 166,000	1881 — 250,000
1868 — 75,000	1875 — 151,618	1882 — 250,000
1869 — 350,000	1876 — 321,998	1883 — 265,000
1870 — 350,000	1877 — 201,000	1884 — 416,000
1871 — 600,000	1878 — 265,649	1885 — 350,000
1872 — 350,000	1879 — 257,927	Total — 5,713,976

[14] Quoted from the *Pleasanton* (Texas) *Journal* in the *National Live Stock Journal*, V, 326, Sept., 1874.

To operate a business on such a basis was indeed placing great confidence in human nature. It might very easily happen that part or all of the cattle gathered would be without any mark whatsoever. Under such circumstances, the drover would be under no necessity of dividing the profits. Thus there appeared at the very outset those difficulties that were never surmounted either by legal enactment or by voluntary association among stockmen. An act of the Texas legislature in 1866 established the principle of the accustomed range. To drive stock from its accustomed range, unless it were possible to prove definitely that such stock belonged to the drover, was declared to be a misdemeanor, punishable by fine and imprisonment. This meant that property rights in unbranded cattle were established by the fact that they ran on a certain range, the usual feeding ground of a certain owner. The plaintiff, in a case arising under this law, would merely have to establish the fact that the cattle in dispute were driven from their accustomed range. Penalties were also provided for branding mavericks with an unrecorded brand or altering any existing brand.[15]

Five years later, when the northern drive had reached considerable proportions, a law was passed requiring that all persons purchasing cattle for driving to market across the northern limits of the state were to brand with a road brand, "a large and plain mark, composed of any mark or device he may choose, which mark shall be branded on the left side of the back behind the shoulder."[16] In the early days of the Texas drives, the enforcement of these laws was full of difficulty for, as has been noted, cattle were in some sections of the state almost common property; indeed, as the northern ranges were soon to realize, many a Texan never fully got over the idea that an unbranded calf was

[15] *Laws of Texas*, 1866, Sess. 11, pp. 187-188.
[16] *Ibid.*, 1871, Sess. 12, p. 119.

anything but common property, to be branded by the first cowboy with an iron handy.

In addition to these legal questions of ownership, which the northern movement and the enhancement of the value of cattle had made serious, there were difficulties beyond the borders of the state. When the herds crossed the Red River, they were in the Indian Territory, and their presence there injected a new element into that much confused and perplexing problem of the final disposition of the Indian. The Plains Indian might be forced upon reservations and compelled to settle down to a civilized way of life, thus ceasing to be a menace to the stock growers. Here in the region north of the Red, however, were Indians well along toward civilization, holding their lands under the most solemn treaties, treaties made to repay them in part for their ruthless removal from their eastern homes when the western frontiersmen demanded their tribal lands. Here in the valleys of the Red, the Arkansas, and the Cimarron, it was hoped that they might remain unmolested and safe. But by that unvarying fate, which seemed to doom the Indian to be located on land that the whites soon found indispensable, they were squarely across the Texan's road to market. The movement of this ever increasing stream of cattle across their lands was the first step in the breakdown of their isolation. This was a far more serious matter than the intrusion of the western immigrant or the occasional gold seeker; the temptation to linger in good grazing country or to zigzag across it in a leisurely manner was too great for most drovers. To the Five Civilized Tribes, pasturage was becoming increasingly important, for much of their wealth was in herds. If unlimited grazing were allowed on their lands, the time would come when their vast grazing commons would be utilized by outsiders, reducing their holdings to only that amount of land sufficient for an agricultural people. When this last stage was reached, the inevitable quarter-section in severalty and the destruction of their tribal unity would be the result.

The Indian Act of 1834 contained a provision penalizing all drovers a dollar a head for cattle driven across any land belonging to any Indian or Indian tribe unless the consent of the latter were first obtained.[17] Thus the Indians had the power legally to debar herds from crossing their territory or to demand a toll. Agents of the more unsettled or less civilized tribes might act for their charges, or, as was the case with the Five Civilized Tribes, the Indians themselves could set the toll and see to its collection. Herds were forced to keep along certain trails, and any deviation or undue lingering was to be regarded as violation of the statute mentioned above.[18] Indian agents were instructed to see to it that these regulations were complied with.[19]

The Five Civilized Tribes were quick to take advantage of this means of protecting their pastures and of raising revenue, and laws were passed by their tribal legislatures levying toll and defining trails. The Cherokee National Council on December 16, 1867, set the toll at ten cents a head for all cattle going through their territory, and the other tribes soon followed suit.[20] The action of these tribes tended to throw the drives further westward beyond their jurisdiction. Here less civilized tribes, such as the Cheyenne and Arapahoe, were being settled in accordance with the

[17] 4 *U. S. Statutes at Large*, 730. The term cattle in the law was interpreted as including sheep. (U. S. *v.* Mattocks; 2 *Sawyer*, 148.)

[18] U. S. *v.* Loving, 34 *Federal*, 715. The doctrine of the permitted trail.

[19] *Regulations of the Indian Department* (Washington, 1884), Sec. 530.

[20] Report of the Senate Committee on Indian Affairs, June 22, 1870, *Sen. Rept.* No. 225, 41 Cong., Sess. 2, pp. 1-3. Reduced in 1875 to five cents a head. *Laws of the Cherokee Nation* (St. Louis, 1875), Chap. XII, Art. I. See also E. E. Dale, "History of the Ranch Cattle Industry in Oklahoma," in the *Annual Report of the American Historical Association*, 1920, pp. 312-314.

concentration policy of the early part of Grant's administration. In this region contacts between Indian and white were for a good part of the Texas drive less regulated and tolls not consistently levied. This shift of the trails affected the cattle dealers along the eastern end of the line of the Kansas Pacific, for it reduced their business, while the newer towns in central and western Kansas profited.[21] By utilizing the Texas Panhandle and the Public Land Strip, drovers might escape these hindrances altogether and the Goodnight Trail, which ran through the Panhandle to Dodge City, became one of the most popular routes.

In addition to these obstacles resulting from the presence of great blocks of Indian country lying between the southern pastures and their market, the Texas cattleman had to face the quarantine laws of Kansas and Missouri. In 1867 both states passed similar laws forbidding the driving of Texas cattle into those states during the summer and fall of each year.[22] Texas fever or Spanish fever was a real danger to the small herds of the Missouri and Kansas farmer. The Texas cattle were apparently more or less immune, but as carriers of the Spanish fever tick, they were a menace to all northern cattle. In the winter the danger was less, and cattle driven north before the first of May were thought to be safe. During the whole period of the open range, the danger from this disease affected in a very marked degree the development of the business, as state and territorial quarantine laws became real barriers to the free movement of Texas stock both to market and to the northern plains.

Finally, there were the westward-moving Kansas farmers,

[21] These tolls resulted in a petition from certain citizens of Kansas presented to the Senate in 1870, praying that cattle dealers be protected in their constitutional rights when driving through the Indian lands. *Senate Journal*, 41 Cong., Sess. 2, p. 647; *Cong. Rec.*, May 13, 1870, p. 3433.

[22] *Laws of Missouri*, 24th Assembly, 1867, p. 130; *Laws of Kansas*, 1867, Sess. 7, pp. 263-267.

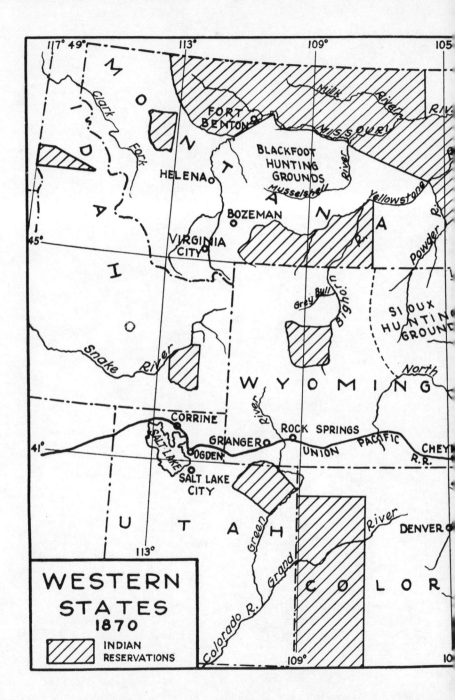

WESTERN
STATES
1870

⬚ INDIAN
RESERVATIONS

whose small enclosed farms soon resulted in continuous lines of fences through which the herds could not pass. The farmers along the western edge of settlement with whom the early Texas drovers came in contact were poor enough fellows, with little inclination to stop the herds. Taking up their quarter-sections in a country of little or no timber, houses and fences were slow in appearing. A dugout cut in the sharp banks of a stream with a roof of cottonwood poles covered with dirt was enough of a habitation to start with, and a furrow turned around the claim was sufficient until fencing could be procured. Fuel was scarce, and early cattlemen were asked by these settlers around Abilene to bed their cattle down for a night or two so that the dung from the cattle might be utilized for the winter's fuel.[23] Later, as the agricultural settlements grew, settlers became less hospitable, and fifteen or twenty Kansas farmers armed with shotguns could argue the most untamed Texas cowboy into going around or paying for the damage done.[24]

Thus the Indian barrier, the quarantine regulations, and the westward-moving farming frontier in Kansas had a tendency to bend the trails out of Texas further and further westward, so that the Texas Panhandle came more and more to be the road to market, and the eastern third of Colorado, the corridor to the northern ranges.

In the early days of the drives, there was very little certainty where the herd would finally find a purchaser. At Abilene, commission merchants from Kansas City might be found and a deal made on the spot. Sometimes good prices were obtained in advance by sending someone ahead of the herd to contract with the buyers. If prices did not seem right along the line of the Kansas Pacific, there was the Union Pacific, three hundred miles further to the north. By slowly

[23] *The Trail Drivers of Texas,* 435.
[24] Letter of C. F. Coffey in *Letters of Old Friends and Members,* 25.

grazing northward, cattle might put on weight and bring better prices at Schuyler, Nebraska, where loading chutes were building and travelers on the Union Pacific were learning to recognize a new type of western settlement, the "cow town." Business was conducted in a primitive enough manner. Kansas City banks handled the paper of the commission merchants, but drovers from central Texas often found it difficult to cash such drafts back home and insisted on cash. Many a herd of longhorns was paid for in gold coins, counted out on a blanket at the campfire of the trail herders. Drafts on the Federal Government issued by Indian agents and quartermasters at the military posts, the first negotiable paper of the Plains, were often used in these early cattle transactions.

The early Texas cattleman could see the possibilities of shipping over the railroads to Kansas City or Chicago; he knew there was a chance to sell his cattle to the farmers of those states where there was a surplus of corn and pasturage; but he had to come in contact with the cattlemen of the north to show him the new market, the northern ranges. The pioneer cattlemen of Wyoming and Montana established contact with Texas along the line of the Union Pacific in Wyoming and Nebraska and in the growing livestock market at Denver. The local market could be supplied with the domestic herds, but if the northern stock grower were to benefit by the new market that the railroad had created and the shipments over the Kansas Pacific had demonstrated as feasible, then he must have cattle to stock the empty plains.

Because of its position, Colorado felt the impact of the Texas invasion first. Along the Arkansas and its tributaries and westward over the Sangre de Cristo range on the Rio Grande, ranch sites were being selected and herds of longhorns were increasing every season. To the north, the herds on the South Platte, which had their origin in the gold rush of the preceding decade, had increased by 1866 to 20,000 head.[25] In that year, an observer saw thousands pastured on the plains in the vicinity of Denver. "Nothing," said he, "short of violence or special legislation can prevent the Plains from continuing to be forever that which under Nature's farming they have ever been — the feeding ground for mighty flocks, the cattle pasture of the world." [26] The possibility that Colorado, the child of a mining boom, might become a great cattle-growing area, the wealth of its herds rivaling the riches of its mines, was beginning to be realized.

The Texas herds did not get into Colorado without some opposition, however, for there was the fear of disease and of glutting the local market. "For several years," remarked the *Denver Tribune* in 1869, "there has been considerable outspoken prejudice in stock-growing and grazing counties against the introduction of Texas cattle into our Territory." [27] In El Paso County, meetings of stock growers were held at which resolutions were passed against Texas cattle, and out on the trails some of the Texas herds were turned backed by groups of armed men.[28] But any prejudice or fear that the early Colorado cattle growers may have had was soon overcome by the opportunities for profit in shipping Texas cattle to the eastern market or in handling them as stockers for the ranges of Idaho, Wyoming, and Montana.

By 1869, a million cattle and two million sheep were reported as grazing within the borders of the Territory. More than half of this number were to be found in the northern area from Denver to the Wyoming boundary.[29]

[25] Report of the Surveyor-General of Colorado in the *Annual Report of the General Land Office*, 1866, p. 102.

[26] C. W. Dilke, *Greater Britain* (London, 1869), I, 136.

[27] Quoted from the *Denver Tribune* in the *Cheyenne Daily Leader*, April 15, 1869.

[28] *Cheyenne Daily Leader*, June 19, 1868.

[29] *Annual Report of the General Land Office*, 1869, p. 153.

Governor McCook, commenting on the great advantages of Colorado in 1870, pointed out that an increase of 60 to 80 per cent in the herds had come about since the arrival of the first Texas cattle. After one season of Colorado grass, he estimated they easily put on a 20 per cent increase in weight. Stock growers further east were coming to Colorado, and eastern capital was arriving for investment in the new industry. Expansion seemed unlimited, for there was still abundant pasturage for thousands of head.[30]

Tributary to the South Platte area were the few towns of flimsy board and flapping canvas into which the Union Pacific had breathed the breath of life and which in 1869 constituted about all there was of the Territory of Wyoming. Before the arrival of the Texas herds and the exploitation of the ranges south of the old emigrant road, these towns were little more than stopping places for transcontinental trains, where locomotives were changed and where passengers got off to stretch their legs, to eat, and to marvel at the great open spaces and the few inhabitants whose doom it was to dwell therein. Life crowded close to the railroad track, and the Union Pacific depot and restaurant were centers of town activity.[31] To the traveler there appeared no reason for towns at all, save to house those who were in some way connected with the operation and maintenance of the road.

There were, however, two other activities that contributed something to the life of these towns. North of the railroad, in the famous old South Pass of emigrant days, gold had been discovered in sufficient quantities to start a mining dis-

[30] Address of Governor McCook before the Colorado Agricultural Society, September 30, 1870, reported in *National Live Stock Journal*, I, 70, November, 1870.

[31] It has been suggested to the writer that a very large proportion of the first pioneer women of Wyoming were waitresses in the Union Pacific restaurants.

trict. Into the new mining camps of South Pass City, Pacific City, and Miners' Delight, flocked a crowd of miners from Colorado and Montana. To these were added gold seekers from the East who could now ride out in comfort and safety on a Union Pacific train. Cheyenne, the largest of the railroad towns and a junction point for the line to Denver, benefited most by this new mining boom. It rapidly became the chief outfitting point for parties bound for the diggings and a center for the freighting business that immediately developed to supply the camps.

The forts along the old Oregon Trail and those that had been established later to protect the railroad furnished the other source of revenue. Those who owned wild hay lands along the river bottoms and who were near enough to the forts to make a profit, furnished forage. The quartermasters were in the market for beef, and those who had cattle to sell got good prices on beef contracts.[32] Finally, the army required the transfer of heavy loads from the railroad to the posts and this furnished some business for the freighters.

Beyond this very limited field, there was nothing to attract settlement or give promise of future prosperity. The expansive enthusiasm, so characteristic of the frontier, did not have quite that note of assurance among the settlers on the High Plains that the certainty of a rapid and complete utilization of the latter by pioneer farmers would have given. "The world is whirling past our doors," wistfully remarked a Cheyenne editor in 1869, "and if the passing multitude behold no signs of agricultural or pastoral life, but instead only a vast expanse of uncultivated and uncared-for land, why, the world will regard us as little else than heathen." [33]

[32] I. W. Iliff from the South Platte had been selling beef in the Wyoming towns ever since their founding. His advertisements occur in the Cheyenne papers as early as May, 1868. *Cheyenne Daily Leader*, May 21, 1868.

[33] *Cheyenne Daily Leader*, May 5, 1869.

The Federal Government had set up the elaborate system of territorial government to rule a few thousand inhabitants along the track. Outside of the railroad itself, the local government found very little taxable property. Indeed, for several years there were indications that Congress might repent of its rashness in setting up a government for a community that was almost nonexistent and turn the region south of the North Platte over to the Territory of Colorado. The rapid development of stock growing in the seventies gave to Wyoming a real territorial status and started it on its growth toward statehood. Thus the history of the Territory and the history of the stock-growing industry therein are so interrelated that the story of one is to a very large degree the story of the other.

To give the exact date of the arrival of the first Texas cattle in Wyoming is impossible. As was noted in the preceding chapter, a herd of Texan cattle had crossed the territory by the old Bozeman Trail in 1866.[34] This was apparently an isolated instance and had no influence on the future development of Wyoming stock growing. In 1868, the bulk of the cattle wintered around Cheyenne were those of the freighters. The local editor, commenting in February, 1868, on the grazing possibilities about the "Magic City of the Plains," ventured the opinion "that the cattle 'upon a thousand hills' around Cheyenne must have the remaining grass pretty well gnawed off by this time from the numbers of them we see made to depend upon this mode of sustaining existence. The grasses of this country may be very nutritious but we would advise a little ration of hay daily as conducive to a more probable certainty of 'fighting it out on that line 'till spring.' "[35]

Among the numbers who were getting through the winter

[34] *Supra,* p. 21.
[35] *Cheyenne Daily Leader,* February 7, 1868.

along with the cattle of the freighters were some longhorns. Iliff had gone down into southern Colorado in the fall of 1867 and there met Goodnight, who had come up over the trail from Texas into New Mexico and thence up along the Pecos into the valley of the Arkansas. Here Iliff bought the herd from Goodnight and in mid-winter, 1867-1868, drove them to the vicinity of Cheyenne, arriving there in February, 1868.[36] In May of that year he was selling a thousand head to the local meat dealers at five cents a pound gross.[37] In April, 1868, the editor of the local paper was advising his readers that the good crop of grass meant plenty of motive power for the freight trains and an abundance of feed on the road to the mines.[38] In an editorial a month later, however, the possibilities of stock raising as one of the most profitable industries of the Territory were being canvassed.[39]

Naturally enough, the Union Pacific became one of the early boosters for the exploitation of the northern ranges. The *Omaha Herald* in 1870 carried a series of letters from a Dr. Latham, at one time employed as surgeon by the Union Pacific. In the next year, these letters were published in pamphlet form and distributed by the railroad all over the country. Dr. Latham's enthusiasm was boundless for the "succulent grasses" so long neglected, which were to nourish a whole half-continent of stock and reduce the price of beef so that no American working man would be so poor as to want for juicy steaks.[40] There began to appear in eastern exchanges an apocryphal story of the belated freighter, storm-bound on the plains, who was forced to turn his oxen

[36] Letter of W. D. Reynolds in *Letters from Old Friends and Members,* 54. Reynolds was a cowhand employed by Goodnight.

[37] *Cheyenne Daily Leader,* May 21, 1868.

[38] *Ibid.,* April 22, 1868.

[39] *Ibid.,* May 8, 1868.

[40] Latham, *op. cit.*

loose on the barren waste and make it on foot to the nearest habitation. When the snow had melted in the spring, he went out with a new team to bring in the abandoned wagon, and to his great surprise, found his old team, sleek and fat, grazing nearby.[41]

Kansas City and Omaha, the two termini of the western roads, were feeling the stimulus resulting from this newly discovered western resource. In 1870 Kansas City banks handled three million dollars of cattle money and a half million passed through the First National Bank of Omaha, as a result of the season's stock business.[42] In 1865, stock interests in Chicago had combined forces and opened the Union Stockyards, little dreaming of the flood of Texas cattle that was soon to overwhelm them. In 1869, these future possibilities were clear, and the directors could report with enthusiasm that in spite of the falling off during the current summer as a result of disease among the Texas herds and the restrictions in Illinois, "the importation of Texas cattle is an immense business and continually growing and with the extension of the railroads through Kansas, and south into Texas, no one can compute the number of cattle in that section that will seek a market in Chicago. The People, and especially the consumers, demand this Texas stock, as it lessens to them the price of beef, and at least for seven months in the year it can safely be brought through the State without detriment to the native herds." [43] As yet, how-

[41] This became the stock story to start off prospectuses for newly formed cattle companies. The grief of the owner at leaving his team or his surprise and joy at finding them alive were needless; any old freighter along the road or any of the pioneer cattlemen along the Platte could have set this "tenderfoot" right.

[42] Quoted from the *Cincinnati Gazette* in the *National Live Stock Journal*, I, 300, May, 1871.

[43] *Third Annual Report of the Directors of the Union Stockyard and Transit Company*, Chicago, 1869.

ever, the future trade from the northern ranges was not great enough to take their eyes off the Southwest.

The stocking of the plains along the line of the Union Pacific may be traced by the location of the chief shipping points of that road. In 1870 the first herds were loaded at Schuyler, sixty miles west of Omaha, where 35,000 head were sold.[44] Later Fort Kearney and North Platte further up the Platte Valley became leading shipping points, while the loading of hogs at Schuyler, showed that the farmer had arrived. By the close of the seventies, Ogallala and Sydney in Nebraska and Pine Bluffs and Rock River in Wyoming were leading.[45] The first shipment of cattle from Wyoming eastward over the Union Pacific has been placed in 1870, when a small herd of work cattle was sold. These cattle, it is said, traveled far, for they helped to feed the French army in the Franco-Prussian War.[46]

The diversion from Kansas City to Omaha of part of the cattle shipments as the herds moved up to the line of the Union Pacific was the beginning of the long struggle between these two cities for the western cattle trade. In 1870, the *Cheyenne Daily Leader* noted that shippers along the Union Pacific had made arrangements with that road and the connecting Iowa roads, the Chicago North Western, the Rock Island, and the Burlington and Quincy, by which rates on livestock were set at figures that would effectually com-

[44] *Fourth Annual Report of the President of the Nebraska State Board of Agriculture*, Lincoln, 1873, p. 58.

[45] *Fifth Annual Report of the Omaha Board of Trade*, 1884, p. 47.

[46] Letter of Hiram B. Kelly, *Letters from Old Friends and Members*, 19-20. The winter shipment of dressed beef from Wyoming appears to antedate that of live animals, for the *Cheyenne Daily Leader* in 1870 noted that I. W. Iliff had for some time been slaughtering beef cattle on Crow Creek (south of Cheyenne) and shipping dressed quarters to Chicago, often one or two carloads a day. *Cheyenne Daily Leader*, January 3, 1870.

pete with those set by the Kansas Pacific and the Hannibal and St. Joe. This latter combination, the *Leader* stated, was struggling to hold the business that, through no effort on their part, had come to them as a result of the Texas drives.[47] The competition among the roads between the river and Chicago was the only circumstance that would have aided the western cattle grower in obtaining reasonable rates on his livestock during the first years of the open range.

The year 1871 was a big year for the Texas drive. Drovers were scouring central Texas for cattle; companies were being formed to finance the drives, and herd after herd was started northward as soon as it could be "put up" and foreman and cowboys hired. One pair of drovers sent ten herds of 1,500 each up the trail in that year.[48] Prices were not as good as in the preceding year, but those quoted in the Denver market reflected the demand created by the northern ranges. Yearlings brought $10; two-year-olds, $15; dry cows, $18 to $20; cows and calves, $21 to $23; and beef steers, ready to be fattened for market by a season or two of northern grass, $25 to $30.[49] Of the 630,-000 head that crossed the Red River in 1871, the ranges of western Nebraska and southern Wyoming got a possible 100,000 head.

There was plenty of room in 1871, and it was not hard to find choice ranch sites with fine water in the nearby streams and a virgin range stretching away from the corrals. In the angle formed by the North Platte and its tributary, the Laramie, was a country of high, rolling, well-grassed bench land and of rich bottoms along the numerous streams where wild hay grew in abundance. Here the early

---

[47] *Cheyenne Daily Leader*, May 24, 1870.
[48] Letter of C. F. Coffey in *Letters from Old Friends and Members*, 25.
[49] *Third Annual Report of the Denver Board of Trade*, 1872, p. 61. Prices were good in the Chicago market in 1870 and 1871.

Wyoming stockmen were selecting locations, making such primitive improvements as were necessary to start operations and stocking the range from the drive. Each of the tributaries of the Laramie and the Platte had herds along its banks by 1871.[50] Westward, over the Laramie Mountains on the Laramie Plains, 15,000 head of acclimated cattle, and 5,000 driven in just at the close of the season, were wintered in 1871-1872.[51] "The season of 1871," comments the *Cheyenne Leader*, "has been a memorable one in the stock business on the plains. Its success was doubted by many newcomers, but the year has closed with their unlimited confidence in the complete practicability and profits of stock growing and winter grazing. The number of cattle is now double, if not four times larger than in 1869."[52]

The Wyoming cattlemen of the seventies were to a large degree men who had lived in the territory before the arrival of the Texas herds. Some of the pioneers who had built up their herds to meet the demand along the overland routes and at the mines were still in the business. They merely enlarged their operations and profited by their previous experience. Keepers of road ranches along the Oregon Trail and owners of hay land, who had been furnishing forage to emigrants and to the military posts, naturally became owners of cattle with unusually good sites from which to

[50] In the report of Silas Reed, Surveyor-General of Wyoming, in the *Annual Report of the General Land Office*, 1871, p. 263, the following figures are given:

Crow Creek	17,700	Box Elder	1,000
Horse Creek	12,800	Sabylle Creek	350
Lodge Pole Cr.	2,200	Chugwater	4,100
North Platte River	14,600	Laramie River	1,800

Before the close of the year, one outfit, Creighton and Alsop, had brought in 45,000 head and turned them loose on Horse Creek. Report of Dr. Latham at the meeting of the Colorado Stock Growers Association at Denver, *Cheyenne Daily Leader*, January 19, 1872.

[51] *Ibid.*

[52] *Cheyenne Daily Leader*, April 11, 1872.

operate. Miners, freighters, and those who had come out on the plains with the railroad got into the business either as small operators or as hired hands. Merchants in the towns, which had declined after the first flush of railroad building had passed away, and storekeepers from defunct mining camps found an opportunity for profitable investment in the steers coming up the trails.

In addition to these local inhabitants, drovers up from Texas with a herd were often induced to stay by the opportunities that the unoccupied ranges presented. Some of these set up for themselves, but many, since the demand for experienced cowhands was almost as great as the demand for steers, hired out to the Wyoming operators. The technique of the range, based partly on the early practice of cattle herding in Texas but to a large degree on the lessons learned in the stress of the northward drives, had to be mastered by these early cattlemen of the northern ranges, and the Texas cowboys were their instructors.[53]

The capital that set these early cattlemen up in business was chiefly local. The East was just beginning to learn of the money-making possibilities of cattle raising on the High Plains, but as yet this advertisement had not reached the

[53] In commenting on the arrival of Texas cattlemen on the Little Missouri in 1881-1882, Lincoln Lang in his *Ranching with Roosevelt* (p. 176) says, "What they [the southern cattlemen] did not know about the open range ranching was inconsequential; although as the future would abundantly prove, even they did not know the Bad Lands. So they took the lead, the rest of us following in line and in due time attaining proficiency in the art under their tutelage."

It is estimated that between 1868 and 1895, 35,000 Texans went over the trails. (*Trail Drivers of Texas*, 453.) With them came the Texas cow ponies, whose contribution to the northern cattleman's education was by no means insignificant. In the early drives, they were driven back to Texas after the cattle had been delivered, but later the demand for Texas riding stock became so great that large herds were driven north each year. In 1884 it was estimated that 100,000 were driven out of Texas. More than a million head came north during the range period. *Ibid.*, 22.

proportions that in the eighties sent millions of Eastern and foreign capital westward. Kansas City and Omaha banks were beginning to handle some cattle paper, and commission merchants in Chicago and elsewhere were furnishing some capital. Out on the plains, money was scarce and interest rates abnormally high — the interest rate at the Cheyenne banks in 1873 was 3 per cent a month.[54] The money obtained by the shipment of stock was immediately turned back into the business, for no investment promised larger returns.

Except for the money to purchase the cattle, investment was slight enough. A homestead entry of 160 acres along some stream was selected as the basis for operations. If there was wild hay on the land, forage for the riding stock was assured. Ranch houses and corrals might come later, but many an early cattleman spent the first season or two in a dugout cut in the hillside near the creek with a similar one close by for sheltering his horses in the worst storms. Beef from his herd, bacon, beans, and coffee, brought by pack horse from the nearest settlement, constituted the bill of fare.[55] As the profits from the first sales were realized, a ranch house and corrals were built. Additional herds were bought and by the third or fourth season, the pioneer cattleman had a going concern. The owner could then leave the ranch in charge of a foreman and move into Cheyenne where a society of cattle owners was in the making.

Cheyenne, which had been absorbed in speculating in town lots, in watching the railroad trains go by, and in outfitting prospectors for the Sweetwater mines, began to become conscious of a larger destiny. In 1871 the local

[54] Letter of C. F. Coffey in *Letters of Old Friends and Members*, 27.

[55] T. H. Sturgis, Secretary of the W.S.G.A., described the experiences and hardships of early stock growing in Wyoming in a speech before the National Stock Growers' Convention in Chicago in 1886, quoted in the *Cheyenne Daily Sun*, November 25, 1886.

newspaper was telling its readers that within a radius of one hundred miles there were between sixty and eighty thousand head of stock.[56] In another year citizens were being solicited for funds to advertise Cheyenne as the cattle market of the plains.[57] Weekly letters from Chicago commission merchants, giving prices at that cattle center, began to appear in the newspapers. During the winter the papers published reports giving the condition of the ranges and of the herds wintering thereon. Texas cowboys lined up at Cheyenne's numerous bars and assisted in making the problem of fostering civic righteousness a difficult one. In celebrating Independence Day in 1872 Cheyenne citizens were diverted by an exhibition of Texas steer riding, and a few months later, a spectacle of bronco busting on Sixteenth Street was regarded as unusual and deplored as cruel by an editor who had not yet learned the ways of the plains.[58] Herds bound for Montana, Idaho, Utah, and Nevada passed through Cheyenne on their way west and it was not long before cattle from as far west as Oregon and Washington were coming into Cheyenne for shipment to Chicago or for sale as stockers.[59] In the fall, Montana steers got down to the railroad in the vicinity of Green River, ready for shipment; further eastward, the stockmen of the Laramie put up their beef herd, and drove to Rock Creek or Laramie; and still further to the east, herds began to appear along the tracks at Cheyenne, Hillsdale, and Pine Bluffs. Into the short, poorly adapted cars furnished by the

---

[56] *Cheyenne Daily Leader*, September 2, 1871.

[57] *Ibid.*, February 24, 1873.

[58] *Cheyenne Daily Leader*, July 6, 1872; September 11, 1873.

[59] The local newspaper notes two herds of transmontane cattle driven to Cheyenne in 1876. One of these, driven over 1,600 miles from the Dalles, was loaded at Cheyenne; the other, from southeastern Oregon, went on to Ogallala before entraining. *Cheyenne Daily Leader,* September 14, 1876; October 27, 1876.

Union Pacific about twenty head were crowded. Strung together in trains of ten to fifteen, these cars jolted to market over the unballasted single track. Freight charges to Chicago were $250.00 a car at Green River in 1879, and $138.00 at Pine Bluffs; so that the Wyoming stockman must deduct anywhere from $6.90 to $12.50 a head from the value of his steers for freight charges in addition to the loss of weight that such a long and rough journey entailed.[60] Some Wyoming stockmen, in order to reduce freight charges, drove further eastward to Ogallala, which in the seventies became the leading cow town in Nebraska. But shipments from Wyoming points were beginning to build up as the following figures show:[61]

	Carloads
1873	286
1874	738
1875	975
1876	1,344
1877	1,649

Since the majority of the Texas drives arrived on the northern range in the vicinity of Ogallala, northern cattlemen in search of stock cattle appeared in this town about the time that the Texas herds were due to arrive. Here the deals between the southern and northern stockmen were made and the herds shifted hands. As time went on and the demand for stock cattle increased, this practice of going out to meet the herds became more and more common, and fewer and fewer herds arrived on the Wyoming ranges in the possession of their Texas owners. One participant in

[60] These figures are obtained from a report made by J. H. Ming to the Montana Stock Growers Association meeting in Helena, January 23, 1879. *Helena Daily Independent,* January 25, 1879.

[61] *Message of Governor Thayer to the Fifth Territorial Legislature,* 1877 (Cheyenne, 1877), p. 10.

this plains cattle market in 1884 gives the following description of it in full swing:

Through the day, the Texas men would work their herds, shape
them up, fill contracts and otherwise employ themselves. The buyers
generally drove out to look at the herds as they came stringing in to
the Platte River. In the afternoon we would meet, have a drink or
two and find out what had been done through the day. We got supper
about six o'clock. The cigars were lighted. Gradually the Texas
crowd would concentrate, go across the tracks, all sit around in a
circle and hold a council of war, smoking, chewing, whittling, com-
paring notes, forming embryo trusts. . . . The buyers did the same,
only they were not so well organized as their southern neighbors.
Then we all met, had confabs together and gradually trades were
made and the cattle began to travel northward.[62]

Although the business conditions of the first part of the
seventies were not good, this process of stocking the available ranges of Wyoming went on unchecked. The panic of
1873 had been hard on the Texas drovers and the prices that
obtained in 1870-1871 in the first flush days of the range
business were not realized again until the boom days
of the eighties.[63] During the seventies Texas cattle were
cheap, and this, combined with the fact that Chicago was
beginning to pay better prices for northern-wintered Texans
and the heavy, well-conditioned cattle from the Montana
ranges, accelerated the expansion of the northern cattle

[62] John Clay, *My Life on the Range* (Chicago, 1924), 111.
[63] The following prices were paid for trail cattle in 1876:

Yearlings	$7.50 to $ 8.50
Two-year-old cows	$12.00
Cows	$14.00
Two-year-old steers	$13.00
Three-year-old steers	$16.00 to $17.00

The above figures were paid for a herd of 12,000 sold to Swan Brothers,
Sturgis, and N. R. Davis. (*Cheyenne Daily Leader*, July 11, 1876.) "The
Centennial Drive" of 1876 was composed chiefly of stock cattle. (*Ibid.*, July
20, 1876.) See also McCoy, pp. 250-251.

business.[64] By 1874 Wyoming had approximately 90,000 head of cattle on its ranges, and each succeeding year showed a marked increase: 110,000 in 1875; 136,000 in 1876; and 176,000 in 1877. With the opening up of the northern areas following the Sioux war, the figures jumped to 260,000 in 1878, 450,000 in 1879, and 530,000 in 1880.[65]

While the herds were increasing every year along the North Platte, and the Wyoming stockman was laying the foundation for future prosperity by buying more and more Texas cattle, the Montana cattle grower, far from the line of the Union Pacific and an outside market, was finding it increasingly difficult to hold on. As has been noted, he had developed herds large enough to meet the needs of the mining camps in the mountains.[66] The increase of these herds in the late sixties was not sufficient to meet the demand for stock cattle that developed as the placers gave out and the mining industry became centered in certain localities. This change in mining methods from the relatively simple process of placering, a process requiring little machinery, to the more complicated methods of deep mining, where machinery was furnished by eastern capital and operated by wage earners living in large settlements, is an economic transformation through which every mining region in the mountains passed. The older crowd of miners scattered, for their experiences had rendered them too independent to settle down as employees of a mining company. Some drifted off to new mining fields, others turned to stock growing and settled down in some likely spot that they had noted in the old prospecting days. Here a cabin and corrals were built and a small band of cattle turned loose on the nearby hillsides. This change from a miner's to a stockman's

[64] *Cheyenne Daily Leader*, May 12, 1875.
[65] *Cheyenne Daily Sun*, January 3, 1884.
[66] *Supra*, pp. 20-22.

frontier created a demand for stock cattle that coincided in point of time with the early Texas drives. In 1871, the local newspaper of Virginia City noted that 800 longhorns were due to arrive in October of that year.[67] In the same year a herd of such cattle was driven northeastward to the Sun River, the frontier of the cattle-growing industry at that time.[68] The only way such cattle could get to Montana was along the line taken by the early cattlemen who drove north from the Mormon danger in 1857-1858.[69] Along this line ran the stage road that connected Montana with the Union Pacific and the rest of the world. Since the road passed through the older stock-growing areas of southwestern Montana, these more settled regions felt the force of this early invasion first, and many of the older stockmen began to move their herds into the less crowded areas to the north and east.[70]

This boom in cattle growing in western Montana was short-lived, however, for the Montana stockman with no railroad at hand soon found his herd increasing far beyond the local demands. In addition to the decline of the mining population, the decrease in the use of oxen for freighting and the substitution of mules and horses shut off another source of profit. In 1874 a local newspaper placed the surplus over and above the local needs at 17,000 head, chiefly four- and five-year-olds.[71] Prices slipped to ten dollars a head for full-grown steers.[72] Cattle roamed at will. Their owners, indifferent as to their condition or increase, simply held on to them with the hope that the building of a railroad

---

[67] The *Montanian* (Virginia City), April 13, 1871.
[68] Kohrs, 1328.
[69] *Supra*, pp. 14-16.
[70] Kohrs, 1328.
[71] The *New Northwest* (Deer Lodge), May 23, 1874.
[72] The *Rocky Mountain Husbandman*, July 10, 1884.

into Montana might justify this expansion of their herds far beyond the demands of the local market.

Among the Montana stockmen there were enterprising ones who were not content to wait for the railroad. Four hundred miles south of them the Union Pacific was making Wyoming stock growing a practical proposition. For the Montana stockmen this way to an outside market was not all that could be desired. It was a sixty-day drive to the railroad at Granger, Wyoming, with the feed poor along the route and the freight charges $112 a car more to Chicago than from Cheyenne, five hundred miles to the east.[73] Nevertheless cattle were shipped. In 1873 a few herds were driven over the range and down to the railroad where they were sold at a profit.[74] In the following year, eastern cattle buyers appeared in Virginia City and Deer Lodge, where cattle were purchased for Chicago at $18.00 to $22.50 a head.[75] These were good figures in 1874, when the decline in prices caused by the panic of 1873 is considered. That Montana stock growers were able to receive such figures, was probably due to the better quality and heavier weight of the native steers. "We marketed some magnificent cattle," Kohrs wrote, "my native three-year-olds usually weighed 1,300 to 1,350 lbs. on the Chicago market, far outweighing a triple-wintered Texan." [76] Some attempts were made to improve the native stock. In 1870, five bulls of pure Durham stock were purchased in Omaha and shipped to Ogden, where they were driven north to Montana.[77] Other shipments of

[73] House Joint Memorial of the Legislature of Montana to the Secretary of the Interior, 1879, *Laws of Montana Territory,* Sess. 11, pp. 131-133. Also report of meeting of the Montana Stock Growers Association, Helena, January 23, 1879, in the *Helena Independent,* January 25, 1879.

[74] The *Montanian,* March 19, 1874; The *New Northwest,* January 1, 1874.

[75] The *New Northwest,* May 29, 1874. See also Stuart, II, 98.

[76] Kohrs, 1328-29.

[77] Stuart, II, 34.

blooded stock followed in the years immediately succeeding.[78]

The center of the stock-raising industry still remained in the Beaverhead and Deer Lodge valleys, where owners found plenty of open land on which to graze their herds. Here in these regions the industry was carried on in a manner far different from that of the range system which was developing in Wyoming and which Montana stock growers learned to employ later in the central and eastern section of the Territory. Using whatever natural barriers there were that would prevent their stock from ranging too far, they fenced in vast tracts of country, which as yet were in no great demand. An observer in the Beaverhead Valley gives the following description of a typical ranch of the period:

This ranch contains 500 acres of land, under fence, mostly meadows with a good dwelling surrounded by a vigorous growth of young cottonwoods. . . . They [the owners] have about 500 head of horses, 3,000 head of cattle and about 3,000 head of sheep, besides a herd of forty shorthorns and seventy-five pure blood Cotswold and Leicester sheep. They also engage extensively in dairying, making 7,000 to 9,000 lbs. of butter each season. They till about a hundred acres of land. Their home ranch [seven miles distant] is well improved. The residence is a kind of frontier log cottage, squatting low to the ground behind a mass of cottonwood and willows. The barn is 50 by 78 feet . . . and is substantially constructed. There is one sheep house here, . . . with walls about six feet high, close and well covered in with straw and also dairy and poultry buildings. Fences are all good and pastures immense. A fence, six miles long running across the valley, connects the steep, rocky ranges on either side. Five miles above this, is another fence from mountain to mountain, forming an enclosure of thirty miles square of 19,200 acres. Through this, there is a dividing fence along the creek about midway of the valley, thus dividing the pastures of fifteen square miles, one for summer and one for winter range. The one for winter use, contains a warm

[78] Kohrs, 1328-29.

spring. . . . Above this enclosure, is a vast tract of country, to which they enjoy the undisputed right, a range that is so hemmed in that stock cannot get out in winter and seldom do in summer.[79]

To a ranch unit similar to this, Montana stock raisers would return after the experience of the open ranges in the eighties.

To the north and east of the Montana settlements lay the country where the range industry was soon to flourish. Newcomers who could not find room in the older valleys and older stock growers who began to find conditions too crowded worked their way out into these areas. The first movement was north to the Sun River along the road that ran by the base of the mountains from Helena to Fort Benton. In 1871 a herd was driven up to the ranges along the Sun, and others soon followed.[80] To the east lay the Smith River, and beyond, the valleys of the Judith and Mussel-shell, a region of well-watered, natural pasture land, perfectly adapted to cattle growing. Into these valleys the frontier cattleman began to penetrate with his herds.

By the early seventies the usefulness of the northern ranges as a stock-growing area had been demonstrated. In Wyoming the plains south of the North Platte were being rapidly stocked as year by year the Texas herds came up over the trails in ever increasing numbers. In Montana the prosperity of the pioneer stockman would be assured as soon as a reasonably accessible market could be obtained. Railroads, observing the profits that this new activity brought to the Union Pacific and the Kansas Pacific, began to think of western extension. But after all, the northern end of the High Plains was a limited area in spite of the tendency to

[79] *Rocky Mountain Husbandman*, April 13, 1876. The total assessed land in 1876 in the county in which this ranch was located was only about 50,000 acres.

[80] Kohrs, 1328; Stuart, II, 98.

talk about any section west of the Missouri in expansive terms. Eastern Colorado and western Kansas might develop for a time unhindered, but north of the Platte and east of the settlements in western Montana lay a great area of country from which all white settlement was excluded. The northern two-thirds of Wyoming and the eastern two-thirds of Montana was still Indian country. This barrier shut off the Montana cattlemen from an easy access to the eastern market and, at the same time, prevented the northern expansion of the cattle growers in Wyoming. The history of the northern end of the cattle range during the seventies is concerned with the movement up to the edge of these areas, the attempts to break down the barriers that such areas imposed, and, when the barriers were removed, the rapid expansion of the herds in Montana eastward and of those in Wyoming northward, and the merging of these two as the cattle industry entered upon the boom period of the eighties.

III

THE INDIAN BARRIER

Eastern passengers on Union Pacific trains in 1869 might readily count on the chance of seeing from the car windows a band of hostile Indians or a herd of buffalo. Any easterner with enough money and courage to buy a railroad ticket might see with his own eyes the untamed West, an experience hitherto vouchsafed only to the man willing to endure the two invariable concommitants of frontier life, hardship and danger. At no time in the history of the American frontier was the contact between civilization and savagery so complete, nor the contrast so sharp as on the High Plains where the last chapter of that story was being written. The railroad destroyed frontier isolation, lessened danger and exposure, and quickened the processes of transforming a wilderness into a settled community.

This increase in the tempo of frontier life was nowhere more apparent than in the rapid solution of the Indian problem. Warlike tribes who had put up a ferocious resistance to the encroachment of whites upon their hunting grounds, who had fought to a standstill the troops sent against them by the Government, were in a few years broken, subdued, stripped of their power, driven from their hunting grounds, and reduced to helpless wards of their conquerors. The western thrust of travel and settlement cut the Indian country in two; the mining booms in the mountains planted active and growing settlements in the rear of the tribes; and the railroads furnished a means of rapid troop movement and supply, increased the rapidity of settlement, and suddenly

exposed the whole structure of Indian society to the distin-
tegrating forces of the white man's civilization.

An immediate solution of the Indian question, however,
seemed far away to the people of Wyoming and Montana
in 1870. Their expansion was limited, their connection with
the outside world was often blocked and their lives and
property were endangered by the presence of huge blocks of
Indian country from which they were rigidly excluded but
which served as bases for constant forays on the settlements.
The pioneer cattleman was particularly affected by these
conditions, for his search for new pastures took him up to
the edge of the forbidden areas, his prosperity depended on
an easy access to eastern markets, and the nature of his
business and of his property exposed him to Indian attack.

In southern Wyoming he might develop the range industry in those regions favorable to that pursuit, but any
expansion north of the North Platte was impossible. Practically all of the country north of the southern boundary of
Dakota and of the North Platte-Sweetwater line, bounded
on the east by the Missouri River and on the west by the
Big Horn Mountains, and extending north to the Yellowstone, was either Indian reservation or Indian hunting
ground. Early in the sixties a road that cut straight across
this region from the line of travel along the Oregon Trail
to the upper Yellowstone had been opened. From here it
led to the mining camps in western Montana. Along this
road, the famous Bozeman Trail, the Sioux under the able
leadership of Red Cloud had fought it out with the forces
of the Federal Government for the control of that ancient
hunting ground of the Indians of the upper Plains region.[1]
The Sioux won; the forts, which had been constructed to
protect the traffic to the mines, were abandoned, and by the

[1] G. A. Hebard and E. A. Brininstool, *The Bozeman Trail* (Cleveland,
1922).

Treaty of 1868, the Government agreed that all the country lying east of the Big Horns should be regarded as the western extension of the Sioux Reservation in Dakota. In this region, in the valleys of the Powder, the Tongue, and the Rosebud, the Sioux were to hunt the buffalo as they had done from time out of mind, and whites were neither to traverse nor to occupy this area.[2] Thus the practice of providing a western hinterland for each reservation, where the Indians could support themselves by the chase, a practice that had been a part of the Indian frontier policy of the thirties, was continued in the case of the Sioux.

West of the Big Horns, in the Big Horn Basin, there was a good grazing region. The risk of Indian incursion through the mountain passes to the east was great enough, however, to keep the cattleman out of this area, for the Government, which was successful in keeping the whites out of the Powder River country, failed to keep the Sioux within the treaty boundaries. At the head of the Big Horn and its tributary, the Wind River, the Shoshones under the patriarchal Washakie had been granted a reservation including some of the best grazing ground of central Wyoming. Thus, for all practical purposes, Wyoming consisted of the southern third of the Territory as shown on the map, the country lying south of the line of the old Oregon Trail.

Limited as the Wyoming stock growers were, they were better off than their Montana neighbors. Although they were prevented from expanding northward, they at least had access to eastern markets. But Montana was one of the least accessible of the far-western territories. From the Union Pacific, the Montana road branched off at Corrine on the Bear River in Utah. This road followed in a general way the trail made by the cattle traders on the Oregon Trail who had brought their herds over it to winter in the upper

[2] *Indian Laws and Treaties of the United States,* II, 1002.

Missouri country. It was 358 miles from Corrine, described in a contemporary railroad folder, as the only Gentile city in Mormon territory, to Virginia City, and another 125 miles to Helena.[3] Daily stages, making the trip in four days, and fast freight and express, covering the distance in nine days, were in operation as soon as the mines opened. Hundreds of wagons and ox-teams bound for the diggings crowded this road as early as 1863.[4] This was the only practical land connection with the East during the first decade of Montana history.

To the north and east of the mining center was Fort Benton, located at the head of navigation on the Missouri River near its junction with the Marias. In the spring, when the Missouri was running banks full, steamers from St. Louis tied up at the wharves of this old trading post. West and south ran the Helena road, which Lieutenant Mullan had laid out in 1855-1856, and along this bull teams dragged the heavy freight that had come up from St. Louis over the cheaper water route.[5]

The natural eastern outlet down the valley of the upper Yellowstone and southward along the Bozeman Trail had been closed, as we have seen, by the Treaty of 1868. A possible outlet west of the Big Horns, along a wagon road laid out by Jim Bridger and known as the Bridger Cut-Off,

[3] *Nelson's Pictorial Guide Book, The Central Pacific Railroad* (New York, *circa* 1875), 9.

[4] Expedition of Capt. James L. Fiske to the Rocky Mountains, 1863, *House Ex. Doc.* No. 45, 38 Cong., Sess. 1, p. 31.

[5] Capt. W. A. Jones, Report upon the Reconnaissance of Northwestern Wyoming Made in the Summer of 1873, *House Ex. Doc.* No. 285, 43 Cong., Sess. 1, p. 57. Jones gives the following freight shipments to Montana, 1871-1873, as follows:

Via Union Pacific	*Via Missouri River*
1871........................7,501,280 pounds	13,000,000 pounds
1872........................6,129,644 pounds	10,000,000 pounds
1873........................6,000,000 (*circa*)	6,000,000 pounds

was barred by the Crow Indian Reservation, which included the whole of the Big Horn Valley lying north of the southern boundary of the Territory. A wagon road was proposed still further west, flanking the Crow barrier along the upper Yellowstone in the newly created Yellowstone Park.[6]

North of the Missouri and its tributary, the Sun River, was the country of the Blackfeet, Piegans, Gros Ventre, and Assiniboines. This enormous block of territory stretched from the Rockies almost to the eastern border of Montana Territory and southwestward so as to include the angle formed by the Missouri and the Yellowstone east of the Musselshell. In addition to this, by the Treaty of 1855, the country west of the Musselshell and south of the Missouri was assigned to the Blackfeet as their hunting ground, and if the Montana stockmen took their herds into this fine grazing area, they must share it with the most warlike of the northern tribes.[7]

As the Montana cattlemen moved up to the edge of the Indian country the people of the Territory became more conscious of the restrictions imposed upon them. Nothing is more common in frontier history than the incessant demands from the West for the reduction of Indian lands. Memorials to Congress from territorial legislatures, complaints to the Indian Commissioner, and demands in the local newspapers all mark a stage in frontier development common to all the territories where the settler found his chance for unlimited expansion checked by the lines of Indian reserves. The first legislative assembly of Montana, meeting in 1864, when Montana was but a few scattered mining camps in the mountains, complained in a memorial to Congress that only a

[6] Letter from the Secretary of War concerning a Military Wagon Road in Wyoming and Montana Territories, *House Ex. Doc.* No. 22, 43 Cong., Sess. 2, pp. 1-3. This latter route, although it raised some hopes at the time, never became practical.

[7] 11, *U. S. Stats.*, 657.

small fragment of the Territory was open to settlement, all
the country of the Missouri south of the forty-seventh
parallel being Indian country.[8] When the arrangements were
being made to set up the Crow Reservation south of the
Yellowstone, which were ratified by the treaty of May 7,
1868, the legislature objected to a surrender that would
"arrest the tide of empire in the Territories."[9]

The most troublesome of the Montana Indians were the
Blackfeet, into whose country the frontier cattlemen from
the south were moving. In 1867 the legislature was urging
Congress to get these Indians onto a reservation.[10] But it was
not until 1873 that they were given, by executive order and
with their assent, a definite reservation, the southwestern
boundary of which was the Sun and Missouri rivers. In the
next year, Congress restricted them still further by moving
the line from the Sun to the Marias River.[11] This restriction,
however, was not enough, and we find another memorial to
Congress in the same year, 1874, setting forth that most of
the land in the Territory south of the Yellowstone belonged
to the Crows and that all north of the Missouri was set aside
for the northern tribes. As to the remaining area between
the two rivers, the memorial complained that it was the proposed route of the moribund Northern Pacific where the
land grant to the railroad had reduced the opportunity of
settlement. This resulted in the people of the Territory feeling "the burden of railroad legislation before they got the
benefits."[12]

The defense of the northern and eastern frontiers of the
Montana settlements against possible raids rested on the
garrisons of the three forts established in the late sixties:

[8] *Laws of M. T.*, 1864, Sess. 1, p. 721.
[9] *Laws of M. T.*, 1867, Sess. 4, pp. 273-279.
[10] *Laws of M. T.*, 1867, Sess. 3, pp. 268-269.
[11] 18 *U. S. Stats.*, 28.
[12] *Laws of M. T.*, 1874, Sess. 8, pp. 177-180.

Fort Shaw on the Sun River, Camp Baker (later Fort Logan) on the Smith River, and Fort Ellis just east of Bozeman. In 1876 territorial newspapers were urging the establishment of another post far to the east on the Musselshell to protect the new stock-growing interests that were growing up there.[13] Such a post, Fort Maginnis, was built in the Musselshell country in 1880.

To the Montana frontiersman, it appeared that this Indian occupation of half of the Territory was likely to last for some time. From the Cheyenne River in Dakota to the upper Yellowstone the Indian held sway, for the wandering miners had not yet discovered gold in the Black Hills, a development that broke down the Indian power at its very center. All this was in the future, and Montana settlers must take what hope they could from the progress of the Northern Pacific, now rejuvenated through the financial activities of Jay Cooke. Work started in 1870, but in 1873, when the railroad had reached Bismarck at the big bend of the Missouri, the failure of the house of Cooke and the panic of that year stopped all further construction. Bismarck was seven hundred miles from Helena and through Indian country most of the way.

Although the people of Montana were handicapped in their development by the barriers that the presence of great Indian reservations imposed, they did not stand in as great danger of losing their lives and property as did their Wyoming neighbors. From the very first, Indian forays from the north down upon the Oregon Trail had been an accepted risk taken by those who traversed it. The murder of lonely freighters and bands of emigrants too weak in numbers to present an adequate defense, the burning of isolated stations and road ranches and the massacre of their inhabitants were common occurrences. The troops that the Government

<hr>

[13] *Helena Daily Herald*, February 11, 1876.

could or would furnish were never sufficient to give more than the most meager protection. The forts established along the Trail were long distances apart and the garrisons small. The treaties that the Government made with the Indians, which, it was hoped, would bring peace to the Trail and safety to the thousands who traversed it, were of little value, for even though the leaders of the tribe might refrain from warlike acts, the younger braves could not be restrained from running off stock, burning, and killing. The failure of the Government to conquer the Sioux in 1866-1867, the abandonment of the country between the Platte and the Yellowstone, and the destruction of the forts along the Bozeman Trail did not tend to lessen the sense of power of a tribe that had met the forces of the Federal Government and had succeeded in turning them back.[14]

The forts established by the Federal Government in Wyoming were located chiefly with an eye to the protection of the lines of east and west communication. The old trading posts, Fort Laramie and Fort Bridger, were purchased by the Government, one in 1849 and the other in 1858, to guard the Oregon Trail. In 1869 Fort Stambaugh was placed at South Pass. When the Overland Trail to the south, which was followed later in a general way by the Union Pacific, was opened, a new fort, Fort Halleck, was built to guard this new road over which the U. S. mail traveled. At the time of the construction of the Union Pacific, forts were established for the defense of that line: Fort D. A. Russell, near Cheyenne, in 1867; Fort Sanders, on the

[14] Gen. C. C. Augur, Commander of the Department of the Platte, writing to Gen. Nichols in 1867, gave the following argument against surrendering the forts:

"To yield to their demands would be regarded by them as evidence of our inability to hold them [the forts] and would, I fear, embolden them to enlarge the sphere of their hostilities and diminish very materially the chances for a permanent peace with them." *Annual Report of the Secretary of War*, 1867, p. 59.

divide west of Cheyenne, in 1866; and Fort Steele, at the crossing of the North Platte, in 1868. To the north the attempt to hold the line of the Bozeman Trail to Montana resulted in the erection of three forts in 1866 along this route: Fort Reno and Fort Phil Kearney in Wyoming and Fort C. F. Smith just over the Montana line. Two other forts along the old trail were built at the same time, Fort Fetterman, 1867, where the Bozeman Trail left the Platte, and Fort Casper, 1866, fifty miles further west, near where the Bridger Cut-Off left the main road. The failure of the Sioux War of 1866-1867 resulted, as we have noted, in the abandonment of the forts along the Bozeman Trail. The military frontier of Wyoming fell back to the North Platte-Sweetwater line, where it remained for a decade. To assist in protecting the mining camps in the vicinity of South Pass, a small post was established in 1869 on a tributary of the Little Popo Agie in the Wind River Valley.

The defense of the stockman's frontier of the seventies rested chiefly on the three forts in the eastern corner of the Territory, Fort Laramie and Fort Fetterman to the north, and Fort D. A. Russell to the south. As soon as the ground was dried out enough in the spring for the Indian ponies, the raids on the stockman along the Chug, the Sybille, and the Laramie began. Stock were run off, ranch houses burned, and herders murdered. Detachments of cavalry from the forts were ordered out, but the damage was usually done and the Indians away to the north before they arrived. Ranch owners were compelled to rely upon whatever defense they themselves could muster, and men, guns, and amunition were sent out from Cheyenne to garrison the ranches.[15] To the west, the mining camps in the South Pass district re-

[15] In the spring and summer of each year the local newspapers were full of stories of Indian raids with appeals for men and arms to protect the ranches and the mines. *Cheyenne Daily Leader*, April 23, 1873; January 1, 1874; March 3, 1874.

ported the murder of small bands of prospectors and of pioneer farmers who were establishing a little farming frontier dependent on the mines in the upper Wind River Valley.[16]

"Our people have been slain by the dozens and their property destroyed to the extent of over one-half a million dollars within the last few years," complained the territorial legislature in 1875, and in characteristic frontier fashion demanded that their words be listened to as well as those of the Indian-loving eastern fanatic.[17]

In addition to demanding that they be relieved of the continual threat of Indian raids, the people of Wyoming joined with those of Montana in urging the reopening of the Bozeman Trail.

"Give us a Custer, a Carr, or a Sheridan, with a strong cavalry arm of the service," was the demand in a memorial to Congress sent by the first Wyoming territorial legislature in 1869, "a general who will open up the old Powder River route and deliver the settlements from the constant danger of Indian attack." [18] We have noted the handicaps that the existing Indian arrangements and the closing of this route had imposed on the Montana people. To the people in the neighborhood of Cheyenne, the surrender of this road destroyed the possibility of that city's becoming an entrepôt for the Montana trade. Rumors that the old road was soon to be opened became a stock story with which the territorial newspapers of Montana and Wyoming peri-

---

[16] A Cheyenne paper in 1868 noted the beginning of agriculture in the Wind River Valley. "The first sod which has ever been turned in the Wind River Valley for agricultural purposes, a circumstance which occurred on the 21st of the present month, was the starting point which is to decide the value of those mountain lands for farming." (*Cheyenne Daily Leader*, May 29, 1868.) Potatoes from this region were selling in the Sweetwater mining camps in 1869. *Ibid.* July 24, 1869.

[17] *Laws of Wyoming Territory*, 1875, Sess. 4, p. 630.

[18] *Laws of Wyoming Territory*, 1869, Sess. 1, p. 731.

odically aroused the hopes of their readers. Such stories were usually followed by news from Washington that the Indian Bureau regarded such a policy as ". . . highly inexpedient and dangerous to the peace of the frontier." [19]

In spite of discouragements from Washington, ambitious schemes were afoot as early as 1870 to build a Cheyenne-Montana railroad. Cheyenne promoters in that year organized the Cheyenne, Iron Mountain, and Pacific Railroad, and in the following year the Wyoming delegate introduced a bill in Congress providing for a Federal grant of land to aid in the building of such a road.[20] The bill was referred to the proper committee, but was not reported out. The fear of Sioux hostilities if such an enterprise was begun and the possibility that the Northern Pacific would be built through Montana, now that Jay Cooke had taken it up, were reasons enough why the scheme received no encouragement in Washington. The importance of this railroad promotion lies in the fact that it served to keep alive the agitation for a change in the Indian situation.

More potent than the Indian raids or the vague talk of railway extension in bringing about the freeing of Wyoming and Montana from Indian control was the belief that in this forbidden region were gold deposits of a value far surpass-

---

[19] *Cheyenne Daily Leader*, February 17, 1874. The sympathy of the eastern newspapers for the Indians was particularly resented by the West. The *Omaha Bee*, because of its fears that the action of the white settlers would precipitate another Indian war, was christened "Red Cloud's Western Organ." The reception of the Sioux chiefs in Washington by Grant was denounced and ridiculed.

[20] *Cheyenne Daily Leader*, December 16, 1870; *Annual Report of the Commissioner of the General Land Office*, 1871, pp. 288-289; *Congressional Record*, 42 Cong., Sess. 2, May 13, 1871, p. 1510.

Not until 1886 did the building of a railroad north from Cheyenne get started. In May of that year the Cheyenne and Northern was organized and plans were made to build to the Montana line via Fort Laramie. This line, which is at present a part of the Colorado and Southern, was built as far north as Wendover, Wyoming. *Poor's Manual*, 1887, p. 849.

ing any that had been discovered up to this time. Since the discovery of gold in the sands of Cherry Creek had demonstrated the possibility of the existence of mineral wealth in the Rocky Mountains as well as in California, every mountain valley from Montana to Arizona into which the miner dared penetrate had been visited, and the track of the prospector was left on the sands of every stream. As the gold played out in one gulch, the news of a new El Dorado set the horde of restless miners scurrying away to the new diggings.

One of the characteristic legends of the Mineral Empire, which appeared in a variety of romantic forms, was that somewhere, in a yet unexplored region, lay fabulously rich deposits, exceeding in abundance and purity anything yet uncovered. The reason for the prevalence of this story and its variations may rest on nothing save the incorrigible optimism of the prospector and the very human love of tales of vast riches. Still, the story usually started off with the words of Father De Smet when he heard news of the discovery of gold in California. Commenting on the wealth of the California mines, he was reputed to have said that during his visits to the missions in the mountains the Indians had told him of gold deposits, and had shown him specimens which had convinced him that the Rocky Mountain region contained riches far beyond anything that California could produce. Given this as a starter, imagination and desire built up a whole cycle of tales. Some said that the wily Jesuit had made an accurate map of these deposits, which he had secretly sent to the head of his order at Rome. Indeed such unquestionable historical authorities as Ned Buntline and James Anthony Froude were cited as evidence of the truth of this part of the story.

There was the tale of a small band of emigrants who had happened upon a stream with sands of gold. They had gone down to the stream to obtain water in some wooden buckets which the story invariably described as blue. Indians attacked them and they fled, leaving these buckets behind them. All were massacred, save one or two of their number who told the story, but neither they nor the hundreds of prospectors who had an eye open for the "Blue Bucket Diggings," were ever able to locate the spot. The dying words of an aged prospector, far from his mountains, in California, or Oregon or St. Louis, who had built a cabin near a lode of unparalleled richness, set the miners talking about the "Lost Cabin Claim." Thus the story, with endless variations, ran through the mountain camps and played its part in the pressure of the mining frontier on the Indian country.[21]

When the mines in Montana and along the Sweetwater in Wyoming began to play out, it was natural, then, for the notion of vast gold deposits in the Indian country to rise to a positive conviction. To penetrate this region was sure to result in clashes with the Indians. It was equally certain that the Government, determined to enforce the Treaty of 1868 and to maintain some kind of peace along the border, would take measures to keep off intruders. But the people of Montana and Wyoming rather welcomed an Indian war, and they were completely contemptuous of the peace policy of the Grant administration, which they laid to the machinations of the "Indian Ring" and the misapplied sympathy of eastern humanitarians. Obviously the thing for them to do was to organize expeditions large enough to fight off any Indian attack, enter the Indian country, find the gold, and then let the Government adjust the treaty to fit *un fait accompli.*

[21] The local papers give most of the varieties of the story. The *Cheyenne Daily Leader,* February 11, 1873, carried a reported interview with Father De Smet at St. Louis on February 7, 1873. See also Strayhorn's *Handbook of Wyoming,* 13-14. Old miners still tell the story of the "Blue Bucket Diggings" and the "Lost Cabin Claim."

Just such an expedition was preparing during the winter of 1869-1870. Cheyenne, not yet conscious of its destiny as the capital of the "cow country," was still thinking of itself as the great center from which future mining operations would radiate into the rich regions to the north. Its leading citizens passed the winter months in organizing the Big Horn Mining Association, which was to go out in the spring and find the gold, wherever it was. An organization was perfected, officers elected, and money for the expedition raised. All prospectors, all adventurous spirits, and all citizens who desired to see their Territory expand and prosper were asked to join in this enterprise, which was to transform Wyoming from the weakest of territories into a great mining state.

When spring came around the expedition was ready to move. Only the vaguest information was available as to its destination, partly because the members of the band were uncertain, and partly because they did not wish to be hampered by the Government. But the Indian Bureau got wind of the scheme and called upon the War Department to protect the Indians and force the whites to observe the treaty. As a result, the commanding officer at Fort D. A. Russell warned the Association that they must keep west of the Big Horn Mountains in their wanderings, no matter what the stories were of gold east of that barrier. With this warning, the company of one hundred and fifty, fully armed and supplied, got off in April, "with the express understanding," commented the local newspaper, "that it cannot claim for its discoveries and settlements the usual government protection. It must waive its claim to the protecting folds of the Stars and Stripes, to uphold which many of the Big Horners have just fought through a bloody rebellion." [22]

Great anxiety was felt in Cheyenne for the safety of the

[22] *Cheyenne Daily Leader*, April 7, 1870.

frontier band, but the Sioux were busy stealing ponies from the Crows and the expedition proceeded on its way unmolested up the valley of the Big Horn. When a troop of cavalry that had been dispatched after them to drive them off the Shoshone Reservation and to turn them back if the danger of a clash between them and the Sioux appeared imminent rode into the camp on the Grey Bull, the expedition was already in the process of breaking up. Prospects for gold were meager, supplies were giving out, and the discipline was none too good. Some of the band made their way north into Montana by way of the Clark's Fork of the Yellowstone, and others were ready to follow the troops south to Cheyenne.[23] No gold had been discovered, and the expedition had not resulted in the slightest relaxation of the Government's determination to keep the whites out of the Sioux hunting grounds. Still the gold must be there. "The key to the richest gold mines in Wyoming or perhaps in the world, it seems is not yet within the grasp of the white man," remarked the local editor who predicted that "the tedious process of Indian extinction must go on for years." [24]

Although the Montana miners were as excited over the possibilities of gold in the Big Horns as were the people in Wyoming, similar organizations to penetrate the Indian country did not appear until after the Northern Pacific ceased building in 1873. Above everything else, the Montana people desired an eastern connection, and in February, 1874, an organization known as the Yellowstone Wagonroad and Prospecting Expedition was formed in Bozeman to bring this about. It was announced that the aim of this expedition was to open a wagon road to the head of navigation of the

[23] Gen. C. C. Augur, commander of the Department of the Platte, in his report to the Secretary of War, October 25, 1870, gives an account of the expedition. Annual Report of the Secretary of War, 1870, *House Ex. Doc.* No. 1, 41 Cong., Sess. 3, Vol. I, Part II, pp. 31-35.

[24] *Cheyenne Daily Leader*, August 23, 1870.

Yellowstone, somewhere in the vicinity of the mouth of the Tongue River, thus making possible a water connection by the Yellowstone and Missouri rivers with the rail head of the Northern Pacific at Bismarck.[25] Supplies and money were furnished by the citizens, and the territorial governor helped the cause to the extent of 10,000 rounds of ammunition. Two cannon, large enough to kill Indians in quantity, were purchased. In February, 1874, the expedition of one hundred forty-six men and five officers and two hundred sixty-nine head of stock got under way. They followed in a general direction the old Bozeman Trail which led them into the Sioux country. They camped in the ruins of Fort Smith and got within forty miles of the abandoned Fort Phil Kearney. The Sioux were out in numbers and after several skirmishes the leaders decided that they were neither strong enough to go further nor to establish a post where they were. So the expedition turned back and took up the trail westward to Bozeman.[26]

These expeditions, barren as they were in immediate results, were important in three particulars. First, they emphasized the local demand for a reduction of the Indian country and the restriction of the Indian to reservations of moderate area; second, they focused the attention of the people of the West on these forbidden areas; and third, they obtained considerable information concerning the character

[25] *Bozeman Avant Courier*, January 24, 1874. Addison M. Quevey, "The Yellowstone Expedition of 1874," *Contributions of the Historical Society of Montana* (Helena, 1876), I, 268-284. In 1863 a small band of miners from the mining camp of Bannock in western Montana made a prospecting trip down the Yellowstone and up the Big Horn and Wind rivers and then down upon the Oregon Trail on the Sweetwater. They found very slight traces of gold. Capt. James Stuart's Journal is found in the *Contributions*, 1876, I, 149-233.

[26] The people of both Wyoming and Montana were tremendously interested in this expedition and the papers of both territories carried long accounts of its progress.

and resources of the country. The tremendous excitement that developed when gold was actually found in this region was merely a culmination of a decade of agitation and effort of which these two expeditions were a part.

While the Yellowstone Expedition was turning back along the river toward the Montana settlements, Custer was moving up the Little Missouri toward the Black Hills. On August 2, 1874, miners brought gold nuggets to the general's headquarters and in a few days the news of gold in the Black Hills set the whole mining population of the Rockies on the move and gave to the country a new mining sensation, the like of which had not been experienced since the days of '49.[27] Here was confirmation a-plenty of all the stories that had been told; here events came to pass as the Big Horn adventurers had hoped they would five years before. The miner was on the ground or coming in ever increasing numbers; the Indian was on the warpath; the Government must act.

Southeastern Wyoming became the center of intense activity. Cheyenne prepared itself to become a second Denver. Hordes of gold seekers from the East, driven westward by hard times and this latest lure of gold, crowded every train. Miners from every corner of the Mineral Empire swarmed the streets. Cheyenne fed them, outfitted them, and started them off in great caravans for the Hills. On the Laramie and the Chug, herds were put up to follow the miners, for beef prices in Deadwood were far better than the Wyoming stockman got at the railroad. Even the Montana stockman found the Black Hills market good enough to make the long and hazardous drive. Hardly had the intervening country been cleared of Indian hostiles before herds

[27] Report of Major General George A. Custer on the Expedition to the Black Hills, *Senate Ex. Doc.* No. 32, 43 Cong., Sess. 2.

began to move down the Yellowstone in response to this demand.[28]

The hostilities that broke out as a result of the invasion of the Indian country, the arrival of the army on the plains, and the subsequent campaign meant that the old Indian arrangement was no more, and that when the final settlement was made, eastern Montana and central Wyoming would no longer be in the hands of hostile tribes. The defeat of Custer, June 25, 1876, was a signal for all the West to demand that this area be cleared and that the Sioux power be irrevocably broken.[29]

The mining stampede and the military campaign that followed served as splendid advertisement for the country of the lower Yellowstone and its tributaries. The valleys of the Tongue, the Powder, the Rosebud, and the Big Horn became something more than names. Interest in their agricultural and stock-growing possibilities was aroused, not only in the West, but throughout the whole country. General Sheridan in his annual report to the Secretary of War, in 1875, pointed out the vast resources of this new country as a stock-growing area as soon as it was cleared of Sioux hostiles.[30] The Montana legislature, a few months after the Custer battle, was memorializing Congress, urging the

[28] Drives were made from Smith River in 1878 to the Black Hills. An oral account of a drive in that year was given to the author by a Montana pioneer.

[29] "We want peace and security for white people in Wyoming and Montana," declared a Cheyenne editor in commenting on the Custer battle, "and we know whereof we speak when we say that that peace can not be brought about with bacon and flour or sugar, coffee and beef." (*Cheyenne Daily Leader*, July 7, 1876.) This was among the mildest of the comments. The policy of the Federal Government was excoriated; the Democratic newspapers declared that the West was sacrificed and western lives endangered in order that the Republican policy of holding large bodies of troops in the South might continue.

[30] Report of Gen. Sheridan in the Annual Report of the Secretary of War, 1875, *House Ex. Doc.* No. I, 44 Cong., Sess. 1, Vol. I, Part II, pp. 57-58.

opening of the old route from Bozeman to Cheyenne, pointing out that settlements were already expanding eastward to the Shields River and the Sweetgrass and that the country eastward contained "many fertile valleys, is a fine stock-growing country, and is supposed to be rich in mineral wealth." [31]

Two centers of activity appeared on the lower Yellowstone as early as 1877 as a result of the military operations in that region. At the junction of the Tongue and Yellowstone rivers a fort was in process of erection. On the opposite bank of the Tongue a collection of tents and flimsy shacks was starting on a lurid career as Miles City. The buffalo hunter in from the plains, where the slaughter of the northern herd was nearly complete; the prospector and gambler, the dive-keeper and inmate from the backwash of the Black Hills gold rush; the trader, the camp follower, and the deserter from the wake of the army — these were laying the foundation of what was to become the metropolis of the eastern Montana cow country. [32] Steamers, carrying supplies for the army, had gotten up to Miles City on the Yellowstone and even further westward to the mouth of the Big Horn, 480 miles from the Missouri. A rumor that an army post might be established at this latter point was sufficient to cause a city to spring up with the name of Big Horn City and a population of fifty permanent residents. [33] Hopes were entertained that if Congress would appropriate a little money for dredging, light steamers might get still further westward to a point within a hundred miles of Bozeman. [34]

[31] *Laws of Montana Territory*, 1877, Sess. 10, p. 435.

[32] By the next year Miles City had developed in civic experience far enough to have its first wave of reform, when the local justice performed a belated marriage ceremony on some one hundred couples.

[33] *Rocky Mountain Husbandman*, June 7, 1877.

[34] *Laws of Montana Territory*, 1876, Sess. 9, pp. 199-202. See also Gen. Sheridan's report to the Secretary of War, 1875, *op. cit.*, p. 58.

The desire to exploit the mineral wealth of this region had been the great force behind all this activity. Other motives, even the desire by western Montana for an eastern connection, had been secondary. But the Black Hills, although rich, were far too limited in wealth for the thousands who crowded there, and the mines of De Smet did not materialize in the Big Horns, in the valleys of the Powder and Tongue, nor anywhere else. Thus the miner began to fade out of the picture, and in his stead came the real exploiter of these areas, the cattleman.

Wyoming stock growers, who had for years felt the threat to their lives and property from Indian raids while they themselves must endure the restraint that the Federal Government had imposed upon them, were getting ready in 1877-1878 to move northward as soon as hostilities ceased. The relief felt in southern Wyoming at this time was well expressed by an observer who wrote:

But the Wyoming of today glows with a new life. Peace has dawned, so suddenly that the long fettered frontier has scarce awakened from its ten years of darkened dreaming. To realize that this grand area of nearly 100,000 square miles, crowded with all the bountiful resources of a coveted empire, is at once and forever emancipated from savage sway, may be easy in quiet New England, but not so where the keys of development have always been carried at the girdle of a hostile possessor. To define the thrill which permeates the frame of the first herdsman who pushes his flocks northward across the Platte River at staunch old Fort Fetterman, and sets his feet firmly upon "Indian ground" might also be a prosy task in the East, but in the valleys of Wyoming it will meet an echoing tingle never to be forgotten.[35]

In the autumn of 1878 the Wyoming territorial governor, returning from a tour of the Big Horn country, "met several

---

[35] R. E. Strayhorn, *The Handbook of Wyoming and Guide to the Black Hills and Big Horn Regions* (Cheyenne, 1877), 20-21.

little parties of adventurous pioneers exploring for good locations with the intention of taking in herds of cattle next spring." [36] In the next year, enough stockmen had arrived in the newly created Johnson County to organize a county government.[37] Here in a region as large as the state of Ohio, ". . . the settler locates in the wildest sections without fear of molestation and the lone explorer wanders over our vast domain with a most comforting and gratifying sense of security." [38]

Similarly the herds of western Montana were being prepared to invade this region from the north. Stockmen were riding down from the Musselshell to look over this new empire of grass. Buffalo hunters were clearing the way for them. "The bottoms," wrote one observer, "are literally sprinkled with the carcasses of dead buffalo. In many places, they lie thick on the ground, fat and meat not yet spoiled, all murdered for their hides which are piled like cordwood all along the way. . . . Probably ten thousand buffalo have been killed in this vicinity this winter (1879-1880). Slaughtering buffalo is a Government measure to subjugate the Indians."[39]

"Eastern Montana is booming," wrote one editor. "The shackles that have bound it in years past have suddenly burst asunder and its latent resources are beginning to be aroused

[36] Report of Governor Hoyt to the Secretary of the Interior, 1878 (Washington, 1878), 40.

[37] Message of Governor Hoyt to the Legislative Assembly, 1879 (Washington, 1879), 34. In 1877, even before the northern country was cleared, two counties, Johnson and Pease, (later Crook) were created by the legislature.

[38] Report of Governor Hoyt to the Secretary of the Interior, 1880 (Washington, 1880), 8.

[39] Stuart, II, 104. It was commonly understood that the Government was interested in the rapid extinction of the buffalo, the basis on which the independent existence of the plains Indian depended. In a debate in Congress in 1874, Representative James A. Garfield stated that the Secretary of the Interior had declared that he would rejoice, so far as the Indian question was concerned, when the last buffalo was exterminated. Cong. Record, 43 Cong., Sess. 1, 1874, pp. 2107-09.

and developed. For several years past, western Montana has led the van. The monied men of Deer Lodge have arrested stock on its march from the Pacific states and the rapid development of the mines in Butte and Phillipsburg and other camps have attracted the muscle and capital of the territory. . . . Eastern Montana has suddenly awakened. . . . Stock is pouring in from every hand; farmers are locating land, and the mountains are alive with prospectors." [40]

All this renewed activity was reflected in a demand by the Montana people for immediate railroad connections. In the period following the suspension of railway building in 1873 there were attempts on the part of the people of that Territory to stimulate construction.[41] Subsidies were proposed to both the Union Pacific and the Northern Pacific to complete a railway connection in Montana, but these were either voted down by the people of the Territory as granting too much to the railroad, or refused by the railroads as offering too slight an inducement. Granger influence was at work among the farmers of Montana and the anti-monopoly feeling, characteristic of this movement, prevented the Territory from paying too high a price for railroad connection.

The revival of general prosperity throughout the country, which coincided with the close of hostilities in the Indian country, made local aid to the railroads unnecessary. The inhabitants of Bismarck and the Northern Pacific officials could forecast future traffic, when, in the fall of 1879, over two thousand head of Montana steers arrived in town from the upper Yellowstone, 700 miles to the west. They were ferried across the Missouri to the rail head and were loaded

[40] *Rocky Mountain Husbandman*, December 4, 1879.

[41] *Laws of Montana Territory*, 1876, Sess. 9, pp. 128-147. Three subsidy acts were passed as follows: (1) Act to aid a North and South Railroad, pp. 128-136; (2) Act to aid a Helena and Fort Benton Railway, pp. 136-138; (3) Act to aid the Northern Pacific Railway, pp. 139-143.

on stock cars for Chicago.[42] "These," said the *St. Paul Pioneer Press*, "are forerunners of an immense cattle trade which the Northern Pacific will have when it reaches the Yellowstone Valley. . . . It is possible that these northern regions beyond the Missouri will vie with the ranches of Texas in their vast herds of cattle." [43] In the same year the Northern Pacific began to lay tracks beyond the Missouri. As the road moved westward, 1880-1881, the herds of this new drive increased, and loading chutes had to be built as soon as the rails were laid. From central Montana to Pine Bluffs on the Union Pacific or Bismarck on the Northern Pacific a steer walked off a hundred pounds.[44] Each mile of track meant better prices for the Montana stockman.

The Union Pacific had been galvanized into life under the ministrations of Jay Gould, and in 1877 began to build from Corrine northward.[45] On March 9, 1880, amid general celebrations in the western towns, the Utah Northern crossed the Montana line. "The days when we can look over our broad domain, a monarch of all we survey, are numbered," remarked an orator of the day.[46]

The process that we have been describing, in which the miner, the soldier, and the railroad builder played the chief roles, was the turning point in the development of the northern range. With the close of the seventies the frontier conditions, which had obtained along the line of the old Ore-

[42] *St. Paul Pioneer Press,* quoted in the *Helena Daily Herald,* September 18, 1879.

[43] *Ibid.*

[44] *Rocky Mountain Husbandman,* November 24, 1881; *Helena Independent,* March 9, 1883. In 1881 the Northern Pacific had crossed the borders of Montana and at Glendive and Kieth in that year 11,000 cattle were loaded. *River Press* (Fort Benton), November 16, 1881.

[45] N. Trottman, *History of the Union Pacific,* 175-188.

[46] *Helena Independent,* March 10, 1880; *Rocky Mountain Husbandman,* March 18, 1880.

gon Trail ever since the track of the first white man crossed
the South Pass, disappeared. The last of those areas that
once included the whole of the High Plains where the Indian lived the wild, roaming life of his forefathers was
opened up.[47]

The last vestiges of the permanent Indian frontier were
passing. No longer could the Indian run the buffalo on the
vast, uninhabited plains; no longer would the Government
deal with any Indian tribe as an independent nation; rather,
by the very force of the circumstances that we have been
describing, the Indian must accept the reservation. On a
June day in 1881 the people of Miles City watched sixteen
hundred Indians loaded on the government steamers for the
Standing Rock Reservation in Dakota. They were the band
under the leadership of Rain-in-the-Face that had finally
been rounded up and herded down to Fort Keough by the
troops. Here they had camped along the Tongue River
preparatory to their embarkation. "For two days and nights,
the Indians and more especially the squaws kept up their
dismal howlings on taking farewell of their beloved homes
and hunting grounds." [48] As the steamers disappeared from
view around the bend in the Yellowstone, where Northern
Pacific railroad gangs were at work, the crowd gathered
that day to watch the spectacle must have sensed that the
old era had come to a close.

[47] Oklahoma still remained, but here conditions were unlike those which
had existed in Montana and Wyoming. Here the reservation system was in
full operation by 1880.

[48] From a manuscript written by D. J. Louck in the possession of the
Wyoming State Historian.

IV

THE CATTLE BOOM

WITH the turn of the decade, the cow country entered upon the boom period of its existence. The barriers were down, the railroads had arrived, the depression of the seventies was a thing of the past. America was feeling, for almost the last time, the stimulus that great areas of unutilized land gave to a society prepared to exploit them. "The immensity of the continent," wrote an English observer some years before, "produces a kind of intoxication: there is moral dram-drinking in the contemplation of the map. No Fourth of July orator can come up to the plain facts contained in the Land Commissioner's report." [1]

"Cotton was once crowned king," exclaimed an eastern livestock journal, "but grass is now. . . . If grass is King, the Rocky Mountain region is its throne and fortunate indeed are those who possess it." [2] The whole world was urged to participate in the good fortune that the latest of El Dorados offered. Here a poor man might grow rich and a rich man might see his capital doubled and trebled in a few years. Here lay "the boundless, gateless, fenceless pastures" of the public domain, covered with grasses, which hundreds of observers had declared to be the most nutritious that livestock ever fed on. With no operating expense save that of a few cowboys, some corrals, and a branding iron, one might transform these leagues of free grass into steers at top prices. Again the frontier was holding out to the people of

[1] C. N. Dilke, *Greater Britain*, I, 105.

[2] *Buffalo Live Stock Journal*, quoted in the *Rocky Mountain Husbandman*, November 25, 1875.

the United States a rapid, attractive, indeed, a romantic road to wealth.

This last frontier received abundant advertisement. The Black Hills excitement and the hostilities of 1876-1877 had focused the attention of the country on this region. Many an easterner who came to find gold remained to raise cattle and wrote back of the wonderful possibilities of this new land, which the campaigns of Custer, Crook, and Miles had made familiar to the people of the East. In 1881, the Northern Pacific arrived on the lower Yellowstone, and it was not long before places that a few months earlier had figured only in the news dispatches of troop movements became familiar names on railroad time tables. Prospectuses, issued by the roads tapping the cattle country, described the enormous resources along their routes and gave detailed statements of the profits that could be made.

With the arrival of the farmer's frontier on the edge of the High Plains, the people of the United States became conscious that there were limits to the agricultural utilization of the public domain, that the old methods of plant and animal husbandry were undergoing a change in these new regions, and that the land laws of the United States were unsuitable to a semi-arid region and must be made to conform to the new environment. The discussion of these new questions, which had been going on all through the seventies, was crystalized in the report of the Public Lands Commission sent out in 1879 to investigate the operation of the land laws in the Far West. The report of the findings of this committee and the evidence which they obtained at hearings held all over the western country added greatly to the general stock of knowledge of a region that had been almost a *terra incognita* but a few years previous.[3]

---

[3] Preliminary Report of the Public Lands Commission, 1879, *House Ex. Doc. No.* 46, 46 Cong., Sess. 2. The final report appeared the following year.

Territorial legislatures voted sums to publish and distribute pamphlets, the authors of which described these undeveloped regions in unrestrained superlatives.[4] Indeed, the frontier was never inclined to check its enthusiasm for the land that it was making its own. Even the smallest territorial newspaper filled up its otherwise meager columns with articles pitched in the most optimistic key. Letters received from eastern and middle-western farmers who contemplated moving west to enter the range business never failed of long and most circumstantial answers, which, as the editor hoped, often got into the eastern exchanges.

Eastern papers, particularly livestock and farm journals, carried an increasing amount of western material: letters from stock growers, accounts of observers who had made a trip to Wyoming or Montana to see for themselves this latest of bonanzas, extracts from the prospectuses of newly organized cattle companies and range notes and reports on the movement, condition, and prices of range cattle. English and Scotch papers, which had followed the development of the range-cattle business almost from the outset, gave it more and more space.[5]

The stories of the vast profits to be made on the range, which these agencies spread all over the country, could not fail to cause a rush of men and capital into these new areas. Middle-western farmers could hardly be blamed for speculating on the possibilities of a move west when they read in their livestock journals such accounts as the following on "How Cattlemen Grow Rich."

"A good sized steer," read an exchange in the *Breeder's*

---

[4] The Montana and Wyoming legislatures both voted money to publish the pamphlets of Strayhorn on their respective territories.

[5] McDonald had been sent out by the *Scotsman* (Edinburgh) in 1874 to report on the livestock industry in America and the dangers which its sudden expansion had for the English stock grower. His book, *Food from the Far West,* was published in 1878.

Gazette in 1883, "when it is fit for the butcher market will bring from $45 to $60. The same animal at its birth was worth but $5.00. He has run on the plains and cropped the grass from the public domain for four or five years, and now, with scarcely any expense to his owner, is worth forty dollars more than when he started on his pilgrimage. A thousand of these animals are kept nearly as cheaply as a single one, so with a thousand as a starter and with an investment of but $5,000 in the start, in four years the stock raiser has made from $40,000 to $45,000. Allow $5,000 for his current expenses which he has been going on and he still has $35,000 and even $45,000 for a net profit. That is all there is of the problem and that is why our cattlemen grow rich." [6] Running a Wisconsin or Illinois farm was rather tame compared with such adventures in the land of easy wealth.

Even for the man without the $5,000 there was a chance. Bill Nye, whose humor in the *Laramie Boomerang* found its way into eastern exchanges, showed how the investment might be pared down to an irreducible minimum. "Three years ago," he wrote, "a guileless tenderfoot came into Wyoming, leading a single Texas steer and carrying a branding iron; now he is the opulent possessor of six hundred head of fine cattle — the ostensible progeny of that one steer." [7]

The cattle in Wyoming in 1879 were far short of the number necessary to stock the new ranges, where the uncropped grass called for cattle and more cattle. Wyoming stockmen were down on the trail in 1879 buying up herds as they came in from the south, for their new ranges along the Powder, Tongue, and upper Cheyenne. Western Wyoming, according to the assessors' lists of the three western

⁶ Quoted from the *Denver Journal of Commerce* in the *Breeder's Gazette,* IV, 421, September 27, 1883.

⁷ Quoted from the *Laramie Boomerang* in the *Rocky Mountain Husbandman,* June 14, 1883.

counties, was greatly understocked.[8] Even in 1880, when the numbers had increased five times, the receiver of the General Land Office at Evanston could testify that the western section of the Territory could sustain fifty times as many cattle as were on it.[9]

In 1879, a hundred thousand longhorns got by the Wyoming buyers and trailed north into Montana where they met the advancing herds of the cattleman, moving out of the mountains.[10] In the southwestern section of that Territory, the influx of homeseekers caused by the railroad connection with the Union Pacific and the opening up of the ranges further east, were working an economic transformation. Winter range was falling off, the valley lands were being brought under private ownership, fences were going up everywhere, land values were rising and with them, taxes. In 1879, local newspapers noted that most of the local herds were being driven out of this section for the ranges of the Musselshell and Yellowstone. Small owners, who were settling down to agriculture, were persuaded to sell off their small herds by the good prices offered by the larger operators, bidding against each other for stock cattle to start a beef bonanza.[11] Through the older towns in the

---

[8] *Report of the Governor of Wyoming to the Secretary of the Interior,* 1886 (Washington, 1886), 41-43. Figures for the number of cattle in each county, 1877-1880, are as follows:

	1877	1878	1879	1880
Laramie	58,101	77,374	97,641	113,466
Albany	10,328	12,358	50,560	46,350
Carbon	5,000	20,168	46,338	72,055
Sweetwater	11,380	13,846	18,419	25,945
Uinta	3,801	5,359	9,117	9,681
Total	88,610	129,105	222,075	267,497

[9] Testimony of the Receiver of the General Land Office at Evanston in the *Preliminary Report of the Public Land Commission,* p. 431.

[10] *Rocky Mountain Husbandman,* August 28, 1879.

[11] *Ibid.,* March 29, 1879; April 29, 1880.

mountains, long lines of cattle passed, headed for the open range.[12] From this time on, the type of farm in the older counties was reduced in size, a ranch where winter feeding and close herding of small bands was the rule.

The Wyoming stock grower had developed the range system of cattle growing because the country in which he operated was not adapted to any other method. The Texas men, the acknowledged experts, with whom he had early come in contact, had taught him the fundamentals. The Montana stock grower, on the other hand, was a novice on the open range. When he began to move eastward into central and eastern Montana, he discovered that his old methods, adapted to the mountain valleys he was leaving, must be discarded; he must learn the technique of the open range. No longer was he in a country of high mountain ranges, steep divides, and rapid mountain streams — natural barriers that kept his cattle within a comparatively small area; here were open plains over which the cattle might wander for hundreds of miles. No longer were there small streams on every hand; here water was scarce and, as one traveler had remarked many years previous, "Those cattle which are . . . to depasture these plains in a future time, must be sound in wind and limb to gather food and water the same day." [13] New conditions, that called for cooperation and adjustment, confronted him on his introduction to the range cattle business. The roundup, the maverick, the chuck-wagon, and the cowboy were institutions and terms common enough to the Wyoming cattleman but unfamiliar to the older Montana stock grower in 1878-1879.

Granville Stuart, looking back on this change, which came so suddenly to Montana, describes as follows:

[12] The *Missoulian* (Missoula), May 25, 1882.
[13] T. J. Farnham, "Travels in the Great Western Prairies" (1839), *Early Western Travels*, XXVIII, 350.

It would be impossible to make people not present on the Montana cattle ranges realize the rapid changes that took place on those ranges in two years. In 1880, the country [central and eastern Montana] was practically uninhabited. One could travel for miles without seeing so much as a trapper's bivouac. Thousands of buffalo darkened the rolling plains. There were deer, elk, wolves and coyotes on every hill and in every ravine and thicket. In the whole territory of Montana there were but 250,000 head of cattle, including dairy cattle and work oxen.

In the fall of 1883, there was not a buffalo remaining on the range, and the antelope, elk, and deer were indeed scarce. In 1880, no one had heard tell of a cowboy in "this niche of the woods" and Charlie Russell had made no pictures of them; but in the fall of 1883, there were 600,000 head of cattle on the range. The cowboy . . . had become an institution.[14]

From 1880-1885, the demand for stock cattle on the northern ranges reached to the extreme limits of the available supply. Although the movement out of the state increased year by year, Texas could not begin to fill the orders of the northern stock growers.[15] From the outset the market for Texas cattle had been a threefold one: first, the northern grazing grounds of Colorado, Dakota, Wyoming, and Montana, where the young stock, the yearlings and two-year-olds, were in greatest demand; second, the eastern stock centers, which each year received the beef steers, driven north to the cattle points along the Kansas Pacific, Santa Fé, and Union Pacific; and third, the feeders of the Middle West, who bought full-grown steers for final fattening on

[14] Stuart, II, 187-88.

[15] "Every day brings news to the effect that the bulk of the cattle on the northern trail are under contract before leaving Texas," notes the *Breeder's Gazette* in commenting upon the tremendous demand in 1884. "It now looks as if there would be virtually no through cattle on the open market at Dodge City and Ogallala this fall." (*Breeder's Gazette*, V, 674, May 1, 1884.) The drive that year to Dodge City alone was given by a Kansas correspondent as 394,227 head. *Breeder's Gazette*, VI, 500, October 12, 1884.

the abundant forage of that region. By the eighties, however, all this was changing.

First of all, the railroads had arrived in Texas and were
now tapping the cattle-growing areas of that state. The
Missouri, Kansas, and Texas, building south from Kansas
City, got to Dennison on the south side of the Red River in
1873.[16] A few years later, this line was extended southward
to Taylor, a few miles north of Austin, where it joined the
International and Great Northern, which came up from
Laredo on the Rio Grande. Thus a through north and south
connection between the Rio Grande and Kansas City was
established. At Fort Worth, this line formed a junction with
the Texas Pacific, which ran westward across the whole
State to El Paso on the Rio Grande. Further to the east,
this latter road connected with the St. Louis, Iron Mountain,
and Southern, which ran to St. Louis by the way of Texar-
kana and Little Rock. Thus it was possible to ship cattle
direct from the range to the great cattle markets, Kansas
City or St. Louis, where they might be sent on to Chicago or
shipped north to Omaha. At Omaha, they could be shipped
westward over the Union Pacific to the northern ranges.
A direct north and south rail connection between Texas and
New Mexico and the northern ranges was available by 1881
when the Santa Fé, building southward from La Junta,
completed connections with the Southern Pacific and Texas
Pacific at Dennison. Cattle could be sent over the Santa Fé
and Denver and Rio Grande to Denver and thence north to
Cheyenne over the Union Pacific.[17]

---

[16] *Poor's Manual,* 1874-1875, p. 683. In 1879 this road shipped 4,219 carloads
of cattle out of Dennison. (*Reports and Statements of the Missouri, Kansas,
and Texas Railroad*, 1879, p. 59.) In 1884, this road, now leased by the
Missouri Pacific, was running stock trains in seven sections daily or about
4,000 head a day. *Breeder's Gazette,* VI, 121, July, 1884.

[17] *Poor's Manual,* 1884. By 1886, this route to the northern ranges was
well established. The *Cheyenne Sun*, in May, 1886, notes that the vanguard

The southern cattle grower, whose market in earlier years had been at the end of a long drive somewhere "north of 36," now found the market brought to him. Drovers seeking cattle for the drive northward to the old shipping points in Kansas and Nebraska were now competing with eastern and northern stockmen prepared to ship by rail. In 1884, rail shipment to northern ranges was somewhat experimental.[18] The drovers, who saw their business threatened, contended that the time saved by shipping by rail was not worth the increased cost; for freight charges were three times as great as the expenses of a drive.[19] Those who urged the newer method pointed out that there was not only a saving of time, thereby getting the cattle on the northern pastures when the spring grass was at its best, but declared that the cattle were in better condition to face the northern winter than those arriving late in the season from the trail, where good grazing was becoming scarcer each year.

In addition to the thousands of longhorns that were being thrown upon the ranges, cattle were moving northward out of Colorado. The testimony taken by the Public Lands Commission in 1879 had shown that the ranges were becoming overstocked and the grass was playing out. Not only were the ranges of Colorado crowded with cattle, but the

of the Texas cattle had arrived by rail when forty-five carloads from El Paso came into town. *Cheyenne Daily Sun*, May 13, 1886.

[18] "The experiments of shipping cattle from the south instead of bringing them by the long and tedious trail route have thus far proven a grand success, and proper accomodations along the lines of railroads for handling stock are all that is needed to have the new mode generally adopted." *Breeder's Gazette*, V, 884, June 5, 1884.

[19] A Kansas paper gives the drover's side of it. The cost of driving from southeastern Texas to Nebraska, a distance of a thousand miles, was $1.00 per head, as over against freight charges of $3.00 a head. If the trail were abandoned, the cattleman would be at the mercy of the railroads. The agitation for a national cattle trail was partly due to the competition of the railroads for the northern cattle shipment. *Kansas City Drover's News*, quoted in the *Breeder's Gazette*, VI, 85, July 7, 1884.

increase in the numbers of sheep had developed a condition where the cattleman must fight for his range or get out of the state.[20]

More significant than the enormous influx of Texas stock and the drift northward from the central ranges was the shipment of young breeding stock and stock steers from the farms of Illinois, Wisconsin, Michigan, Iowa, and Missouri. During the seventies, the farmers of these regions had taken up numbers of western range cattle to fatten on the cheap forage of which they had a surplus. Even as late as 1879, the directors of the Union Pacific could report that "a considerable portion of the cattle arriving at Omaha and Council Bluffs are distributed through the rich farming regions of eastern Nebraska and of Iowa. Here the farmers find greater profit in feeding the corn through the winter than in marketing it. . . . The business of thus feeding corn has more than doubled during the last year."[21] During the eighties, when the scramble to stock the ranges was at its height, the process was reversed and cattle moved westward in increasing numbers.[22] Feeders found it more profitable to sell to the western cattlemen at the high prices then prevailing, than to buy stock cattle to fatten for the market.[23]

Eastern stock, commonly known on the ranges as "pilgrims," "states' cattle," or "barnyard stock" were crowded into the western-bound stock trains. In 1882-1884, there

[20] *Report of the Public Lands Commission,* p. 295 *et seq.*

[21] Annual Report of the Government Directors of the Union Pacific Railroad, 1885-1886, *Sen. Ex. Doc.,* No. 69, 49 Cong., Sess. 1, p. 156.

[22] Clay, 66.

[23] The *Barber County Index* pointed to the high prices for stock cattle in 1883 when feeders were losing money. This paper estimated that 80,000 yearlings and two-year-olds had been sent to the western states and territories in 1883. *Barber County* (Kansas) *Index,* quoted in the *Breeder's Gazette,* IV, 73, July 19, 1883.

were as many cattle shipped west as east.[24] In 1884, the newly opened Northern Pacific brought 98,219 head of these "pilgrims" and took out 75,000 head for the Chicago market.[25] In one week, 12,800 head were unloaded at Glendive, already calling itself the "Queen City of the Cow Land." [26]

The throwing of thousands of these young cattle into the newly opened areas increased the risks of the range business tremendously. In the first place, they were more valuable than the Texas longhorns.[27] Second, these eastern cattle

[24] Joseph Nimmo, *The Range and Cattle Ranch Business of the United States* (Washington, 1885), p. 182.

The report of a livestock commission firm in Kansas City which appeared in a stock journal in 1883 will give some idea of the changes in the cattle trade produced by the stocking of the northern ranges. The drive from Texas for that year (1883) was estimated at 260,000 head, practically all of which was contracted for at the following prices: Yearlings, $15.00-$16.50; two-year-olds, $19.00-$22.00; three-year-olds and cows, $24.00-$28.00; cows with calves, $30.00-$35.00.

The westward movement of cattle, mostly one- and two-year-old heifers, was given by the *Breeder's Gazette* as follows:

	Head		Head
Missouri	55,000	Louisiana	10,000
Eastern Kansas	20,000	Mississippi	10,000
Iowa	25,000	Tennessee	5,000
Minnesota	15,000	Florida	15,000
Eastern Nebraska	10,000	Illinois	5,000
Arkansas	15,000		
		Total	185,000

These cattle sold at the following figures: Yearlings, $17.00-$21.00; two-year-olds, $23.00-$25.00; young dry cows, $30.00-$35.00; stock cattle, $30.00-$35.00.

In addition to this, Oregon and Washington had sent into Montana, Wyoming, and Dakota 20,000 head and had received from the east about the same number of high-grade breeding stock. Thus some 500,000 young cattle had changed hands as a result of the demand of the northern ranges. *Breeder's Gazette*, IV, 297, September 6, 1883.

[25] *Rocky Mountain Husbandman*, January 15, 1884.

[26] *Glendive Times*, April 26, 1884.

[27] Judge J. N. Carey stated before the National Cattle Growers Associa-

were less able to stand the rigors of a northern winter, so that the danger of winter losses was correspondingly greater. Third, their importation might lead to the introduction of pleuro-pneumonia, which was very serious in the East during the eighties. If this disease once got started in the West, the whole range industry would be shattered. Finally, these cattle, pouring in by rail, would overstock the ranges almost as soon as they were opened up. The fact that millions were invested in the cattle business, in the face of risks such as these, shows the point that had been reached in this wild speculation in cattle. Old cattlemen, whose experience with the northern winter should have made them cautious, plunged as madly as the newcomers.

If the risks were terrific, the possibilities for making a fortune seemed to justify taking them. Prices were soaring. "Untold millions in Europe and America are ready to invest in range cattle," declared a Colorado livestock journal, "and . . . cattle is one of those investments men cannot pay too much for, since, if left alone, they will multiply, replenish and, grow out of a bad bargain." [28] In May, 1882, the price for beef cattle on the Chicago market reached a level higher than any since 1870.[29] On May twenty-fourth top prices touched $9.35 a hundred.[30] Steers, ready for fattening on the northern range, brought thirty-seven, forty, and even fifty

tion in St. Louis in 1884, that during the preceding year, members of the Wyoming Stock Growers Association had purchased $2,000,000 worth of livestock from Illinois and Missouri, mostly thoroughbred bulls. *Proceedings of the Convention of the National Cattle Growers Association* (St. Louis, 1884), 71.

[28] *Colorado Live Stock Record*, quoted in the *Breeder's Gazette*, V, 998, June 26, 1884.

[29] For prices of beef cattle on the Chicago market from 1864 to 1890, see page 95.

[30] Testimony Taken by the Select Committee of the U. S. Senate on the Transportation and Sale of Meat Products (Vest Report), 1888, *Sen. Rept.* No. 829, 51 Cong., Sess. 1, p. 221.

dollars in Texas,[31] and these might be sold to a Wyoming speculator for sixty dollars.[32] The stocking of the Cherokee Strip in the Indian Territory created a still further demand and consequently an enhanced price.

Stories of vast profits appeared to justify the statement that stock cattle at top prices "would grow out of a bad bar-

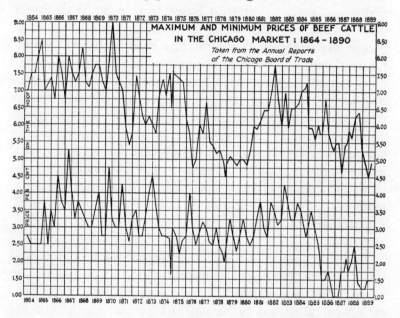

MAXIMUM AND MINIMUM PRICES OF BEEF CATTLE IN THE CHICAGO MARKET : 1864 - 1890

Taken from the Annual Reports of the Chicago Board of Trade

gain." The story was about that the Prairie Cattle Company had announced a dividend of 42 per cent for 1882.[33] Montana stock growers were, according to the livestock journals, reaping a profit of from 25 to 40 per cent.[34] The eastern papers created a new frontier figure, the "bovine king" or

[31] *Ibid.*, 4.

[32] Quoted from the *Kansas Cowboy* in the *Breeder's Gazette*, VI, 50, July 10, 1884.

[33] *Colorado Live Stock Record*, quoted in the *Breeder's Gazette*, V, 313, February 28, 1884.

[34] *Breeder's Gazette*, V, 511, April 3, 1884.

"cattle baron." In the spring, at the head of his picturesque retainers, the cowboys, he rode over the spring range to count his cattle by the hundred thousand, and in the fall was recognized in the Chicago and New York hotels by his Stetson felt and diamond shirt studs. Cheyenne, the center of the northern cattle business, went wild; its streets crowded with speculators. "Sixteenth Street is a young Wall Street. Millions are talked of as lightly as nickles and all kinds of people are dabbling in steers. The chief justice of the Supreme Court has recently succumbed to the contagion and gone out to purchase a $40,000 herd," says the *Laramie Boomerang* in August, 1882. "Large transactions are made every day in which the buyer does not see a hoof of his purchase and very likely does not use more than one half of the purchase money in the trade before he has sold and made an enormous margin in the deal. . . . A Cheyenne man who don't pretend to know a maverick from a mandamus has made a neat little margin of $15,000 this summer in small transactions and hasn't seen a cow yet that he has bought and sold." [35]

Eastern conservatism could not withstand such stories as these, and eastern capital came out on the High Plains with a rush. The pioneer stock grower who had built up a herd on a good range was in the way of becoming a rich man. He might sell out his herd and his range rights at his own figure, or he might exchange his cattle and holdings for capital stock in a newly organized company that would furnish plenty of additional capital for further expansion. Old cattlemen who had kept the calf tally on a shingle, whose only book had been a checkbook, and who knew the state of their affairs only by the balance or overdraft at the bank, found themselves sitting in at directors' meetings where eastern

[35] Quoted from the *Laramie Boomerang* in the *Miles City Daily Press*, August 2, 1882.

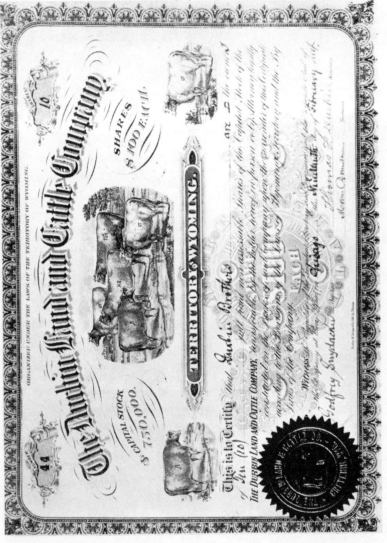

STOCK CERTIFICATE OF A WYOMING CATTLE COMPANY

capitalists deferred to their business experience and judgment.[36] It was all so simple. The United States furnished the grass; the East, the capital; and the western stockman, the experience.

Cattle companies multiplied. In one year, 1883, twenty companies with a total capitalization of over twelve million were incorporated under the territorial laws of Wyoming.[37] Deals involving thousands of head of cattle and

[36] Commenting on the change that had come so suddenly to the ranges, Thomas Sturgis in an address before the Wyoming Stock Growers' Association in 1884 said: "The time has come when our business can no longer be done by the old rule-of-thumb method. In former days we had only to brand our calves, when dropped, and ship our beeves, when fat. The calf tally could be kept on a shingle and the checkbook was the only book kept, and the balance or the overdraft at the bank showed the whole of the business. Times are changed." Clay, 245.

[37] *The Cheyenne Daily Sun*, March 9, 1884, gives the following list for 1883:

Keystone Cattle Co. ....$	500,000
Stodard & Howard Live Stock Co.	550,000
Anglo-American Cattle Co. increased to	800,000
Union Cattle Co.	3,000,000
North American Cattle Co.	1,000,000
Webels Live Stock Co.	10,000
Snow Cattle Co.	500,000
Home Cattle Co.	250,000
War Bonnet Live Stock Co.	300,000
Luke Vorhees Cattle Co.	500,000
Searight Cattle Co.	1,000,000
Belle Fourche Cattle Co.	50,000
Wyoming Hereford Association	500,000
Snake River Cattle Co.	200,000
British American Livestock Co.	250,000
Warren Live Stock Co.	500,000
Northwestern Live Stock Co.	200,000
Wyoming Land and Cattle Co.	250,000
Horse Creek Land and Cattle Co.	200,000
	$10,560,000
Swan Land and Cattle Co.	3,000,000
	$13,560,000

millions of dollars were common. In 1883, for example, the Swan Land and Cattle Company was organized through combining three ranch properties with a range from Fort Steele, Wyoming, on the west, to Ogallala, Nebraska, on the east, and from the Union Pacific to the Platte River. These companies sold to the new company some 30,000 acres of land, part in full title and part in process of title; over a hundred thousand head of stock and rights on a range, one hundred miles long and fifty to one hundred miles wide, for a purchase price of $2,588,825. The new company was capitalized at $3,000,000, which was raised to $3,750,000 to provide for the purchase of over half a million acres of Union Pacific alternate sections.[38]

Part of the capital that went into these new companies came from abroad. During the seventies, Europe became painfully aware of the arrival of the wheat farmer on the hitherto unutilized lands of the Middle West, which the rapid extension of railroads following the Civil War had made available. This and the development of improved agricultural methods and machinery had resulted in an enormous increase in our wheat export.[39] The British farmer with his high-priced land could not possibly compete with the western wheat farmer, and from the middle of the decade, the area in England devoted to wheat raising contracted.[40] Had the English farmer been able to utilize this surplus land by

[38] *Cheyenne Daily Leader*, February 24, 1884; *Wheatland Times*, April 15, 1925; Clay, 202; Coutant Notes in the office of the Wyoming State-Historian.

[39] The effect on European producers of the arrival of cheap American wheat and meat has been treated in a very valuable study by William Trimble, "The Historical Aspects of the Surplus Food Production of the United States, 1862-1902," in the *Annual Report of the American Historical Association*, 1918, I, 221-239.

[40] Between 1878-1902 the area devoted to wheat in England decreased from 3,000,000 to 1,500,000 acres. *Ibid.*, 231.

turning to cattle raising, the effect of the lowering of the price of wheat would not have been so severe. But the utilization of the High Plains as a cheap cattle-growing area and the arrival of increasing quantities of American dressed meat and live cattle cut off this avenue of relief. The practice of shipping live cattle across the Atlantic began as early as 1868. During the next decade the export figures on livestock mounted rapidly.[41] At the same time the development of refrigerating methods in the transporting of dressed meat brought the products of the American packers into the English market to compete with the native beef. Prices of cattle in England fell off. During the period from 1878-1895, the decline in the inferior grades, where the competition was the sharpest, was 40 per cent; in the medium grades, 27; and in the best grades, 24 per cent.[42]

There was nothing new in the investment of foreign capital in far-western enterprises. Western railroads had sold large blocks of securities to English, German, and Dutch investors. Land companies, financed abroad, had purchased millions of acres of the public domain, which had either been settled through some colonization scheme, or were being held as a speculation.[43] The arrival of cheap American beef in the European market and the stories of

[41] The following figures of live cattle export are given for the period, 1870-1879:

Head		Head	
1870	27,530	1875	57,211
1871	20,530	1876	51,593
1872	28,033	1877	50,001
1873	35,455	1878	80,040
1874	56,067	1879	136,720

Annual Report of the N. Y. Produce Exchange, 1879, p. 505.

[42] Trimble, 232.

[43] It was estimated that over twenty million acres were in the hands of foreigners. Texas, Kansas, Colorado, Arkansas, New Mexico, Mississippi, Wisconsin, West Virginia, and Florida were cited as states where foreigners had large holdings. *House Report* No. 3455, 49 Cong., Sess. 1, 1885, p. 2.

the profits that could be made in its production turned the attention of investors to this new field. The *London Economist* in an article headed, "The Latest Development of the Land Company," described the manner in which English and Scotch capital was drawn into the range-cattle industry. After commenting on the fact that, in the years just prior to 1883, the land and mortgage companies in Canada, United States, and Australia had been overdone and that the prospects for success in a land company, pure and simple, were not what they had been, the article went on to say:

It has therefore become necessary to invest such undertakings with an appearance of novelty; and the remarkable success of a Texas cattle-breeding company [Prairie Land and Cattle Co.] started in 1880, which paid a dividend at the rate of 19½ per cent twelve months ago, and had recently declared a second, at nearly 28 per cent, has now supplied the necessary catchword. Every Land Company, now brought forward, must be a Cattle Company as well.[44]

The company mentioned in the preceding quotation, had been organized in Edinburgh and its success became a stock story in the prospectuses of later cattle companies, organized on both sides of the Atlantic. That Edinburgh became the center of this speculation in western ranches, was probably due to the fact that the Prairie Land and Cattle Company was organized there.[45]

It was of little use for the more conservative to point out that such sudden and enormous profits were due to a purely speculative boom and that future investments could not hope to realize such dividends.[46] Scotch investors, with visions of

[44] *London Economist*, XLI, 131, Feb. 3, 1883.

[45] "Had the Prairie Land Company been a London instead of an Edinburgh concern, we should have had the nucleus of the speculation here." *Ibid.*, 232.

[46] The chairman of the Board of Directors of the Texas Land and Cattle Company admitted in a letter published in the *London Economist* that the issue of new shares of stock, "made it imperative to divide among the old

30 to 40 per cent dividends, succumbed to the fever that this "minor South Sea Bubble" had created. "In Edinburgh, the ranch pot was boiling over," writes John Clay, himself an agent for English investors, sent out to report on the prospects in Wyoming. "Drawing rooms buzzed with the stories of this last of bonanzas; staid old gentlemen, who scarcely knew the difference between a steer and a heifer, discussed it over their port and nuts." [47] Big game hunters, filled with enthusiasm for "the great open spaces," could get conservative English bankers worked up to a point where investment seemed a sure thing instead of "betting against God Almighty and a sub-Arctic winter." Glowing prospectuses whipped up the craze for cattle ranching and younger sons of wealthy families found it not so difficult to pry open the family purse for such a profitable investment. Western ranch owners could hardly be blamed for unloading at fancy prices and on "book count" to such investors, who envisioned vast profits from Uncle Sam's free pasture.[48]

All the evils that attend an era of unrestrained speculation and inflation appeared. Paper companies with nothing but a well-written prospectus and attractive stock certificates were organized to catch the unwary. Western papers sent out warnings against such concerns, which were working in the eastern states.

"Although there is not the wild .... vision of wealth connected with livestock as there is in mining schemes, still there is enough in it, ... to induce a great many people of the East to invest money on anything that happens to come their way with a brand on it," warned the *Denver News* in 1884.

shareholders the increment of value on the stock of cattle during the year." *Ibid.*, 233.

[47] Clay, 27-37, 158-187; Stuart, II, 150.

[48] See table p. 102, taken from the *London Economist*, giving the extent and character of English and Scotch investment and the dividends paid in the boom years.

FINANCIAL SET-UP OF ELEVEN ENGLISH AND SCOTCH CATTLE COMPANIES OPERATING IN THE UNITED STATES — 1883-1885

(Taken from the *London Economist*, March 20, 1886, p. 365)

COMPANY	LAND		TOTAL HERD	CAPITAL				ANNUAL DIVIDENDS IN PER CENT		
	Owned Acres	Leased Acres		Debenture Loans and Preferred Stock	Ordinary Share Capital	Shares Amt.	Shares Paid	1885	1884	1883
Prairie	156,862	32,278	124,212	£291,767	£294,055	£10	£5	10	10	20½
Swan	578,853		123,460	288,340	461,660	10	6	6	10	9
Texas	388,174	520,966	106,322	240,000	240,000	10	5	5	6	12½
Matador	424,296	256,367	94,441	200,000	300,000	10	6	7	6	8
Hansford	76,640	124,000	37,734	17,594	209,740	5	5		7	6
Arkansas	16,023	6,080	24,315	124,994	125,000	10	5		6	10
Pastoral	300,692	280,891	45,885	164,289	172,050	10	5	5	8	
Powder River			48,625	100,000	200,000	5	5		4	6
Western Land	75,343		35,469	100,000	100,000	100	5	10*	15	15
Cattle Ranche		153,607	13,500	5,600	100,000	5	5	4	15	5
Western Ranche			18,050	100,000	112,000	5	5	4	7†	

* For thirteen months.

† Dividend of 7% per annum, spread over twenty-two months.

"The purchaser discovers, of course after it is too late, that
the company never existed, except on paper, and that the
great and unreasonable dividends in the prospectus were but
a clever bait to catch the suckers. . . . . There are numbers of
these snide affairs in Colorado and it would be well for the
legitimate companies . . . . to run them down and compel
them to cease their nefarious and robbing business, as it is
bringing discredit on legitimate concerns." [49]

Even those companies with established reputation and
cattle on the range were not above turning off each year a
larger beef herd than the calf crop warranted, thus reducing
the capital in order to distribute a larger yearly dividend.
Companies bought cattle on the range according to the number given on the books of the seller. No tally save the "book
count" was asked or given, for there was no time to go out
on the range and check up the actual number, which might
be scattered over a region scores of miles in extent.[50]

The increase in the number of companies composed of
eastern or foreign investors resulted in all the abuses com-

[49] *Denver News,* quoted in the *Breeder's Gazette,* V, 797, May 22, 1884.

[50] Clay states that in order to pay a 20½% dividend in 1883, the Prairie
Cattle Co. had to sell 21,448 head. Their calf brand for that year turned in
28,207 head. This allowed for a margin of only 7,000 head for winter losses
on a herd of over a hundred thousand. This meant that the company was
dangerously near turning off more than their increase (Clay, 131-132).
That such dividends could be paid by the new companies that had just gone
into the business was out of the question. In commenting on this, the *London
Economist* cautiously said: "These six companies [six of the older ones
organized before 1883] are apparently earning profits; some of them, from
the reports, appear to be in the possession of properties worth more than
they gave for them. It is therefore proved by experience that a well managed
cattle and land, or ranch undertaking can be conducted profitably by a
British joint stock company. . . . But it must not be forgotten that those few
companies which, like the Prairie Cattle Co., bought their estate and their
livestock before the great rise in Western prices had attained its maximum,
have a distinct advantage. The wild prophecies that turned the heads of our
Scotch neighbors twelve months back have justly incurred censure, more
especially as beef continues dear in the Eastern States, and prices are

mon to absentee ownership. Managers, with little at stake, were careless of the condition of the herd. Cowboys wore down the riding stock by careless or abusive handling. Ranch supplies were wasted. Cowhands, out of work during the winter, gravitated to ranches where a reputation for generous hospitality had been built up at the expense of an owner thousands of miles away. The calf crop was far below what the prospectuses had promised. Small bands of cattle belonging to cowboys seemed to grow beyond the natural rate of increase. Somehow, a calf belonging to a company with offices in Edinburgh or New York did not seem quite as much a piece of private property as that of a neighbor on the range. The frontier, which had never taken kindly to absentee ownership, was not so very careful of the property of owners who found it hard to visualize the conditions as they actually existed on the Laramie Plains, the Powder, and the Yellowstone.[51]

This flow of capital out to the High Plains rapidly brought the number of cattle up to the carrying capacity of the ranges. This created a condition that increased the costs of operation and multiplied the problems confronting the stock grower. Investors had been told that once the herd was bought and turned out on the range, nothing remained but to count the profits. But as herd after herd crowded upon those areas that a year or two before had been the exclusive range of some early cattleman, "free grass" was no longer safe. Prudent operators, who dared not take the risk of depending solely upon the crowded ranges around them,

stimulated by a demand in this country; but, like all such foreign or colonial undertakings, a careful first selection and trustworthy and economical management are essential to prevent collapse. And these essentials are difficult to obtain where property covers a vast area in the wilds of a distant country." *London Economist*, XLII, 197, Feb. 16, 1884.

[51] Clay, 54, 158-167; Stuart, II, 50; Vest Report, Testimony, 221 ff; Rollins, *The Cowboy*, 308-309.

began to secure some sort of title to as much of the land as they could. This meant an additional outlay of money, not only for the purchase of land, but also for the construction and maintenance of fences. Others, rather than pare down their profits by increased expenditures, moved their cattle out into those regions that had remained unoccupied, either because of their low quality, or their inaccessibility. Thus, every possible square mile of range that could support an animal was quickly covered. With a rapidity that could almost be measured in months rather than years, every available bit of range in northern and central Wyoming was occupied; the country in eastern Montana, north of the Yellowstone to the southern boundary of the Indian reservation, was filled up, and herds began to look for favorable locations beyond the international boundary along the Saskatchewan River.[52]

Caught between the upper and nether millstones of crowded ranges and increasing expenses, the cattlemen were forced to turn off the matured animals as rapidly as possible. Cattle must go to market; to hold them for better prices became more and more a gamble that endangered the whole investment. At the very time when the cattle boom was at its zenith, and cattlemen were bidding against each other for young stock and she cattle, the prices for beef cattle were beginning to sag. The peak of prices on the Chicago market had been reached in the summer of 1882, when top prices reached a level higher than any figure since 1870. Prices for the lower grades lagged somewhat, the maximum being reached in the following spring. From then on, the decline, particularly in the lower grades, was steady, from $4.25 cwt. in April 1883 to $1.00 cwt., in the winter of 1887.

[52] The old ratio between cattle and population which had obtained before the war was restored. According to the Vest Report (p. 8) there were in 1884, 800 cattle to every 1,000 population; in 1860 there had been 815; in 1870, 618. These figures had been obtained from the Statistician of the Department of Agriculture.

The cutting down of profits, due to the increasing cost of production and the decline in prices, resulted in a closer scrutiny of the costs of transportation and marketing. While the range business was still at its height, and the cattleman still believed he was getting rich, the feeling grew that his future prosperity was coming more and more to depend upon the railroads and the combination of packers and commission men in Chicago. Then came the deflation, caused by speculation, extravagance, and severe winters. As we shall see in a succeeding chapter, there was a real recession of the cattle frontier, as the thousands of head, which had been pushed out on the northern ranges and had been lucky enough to survive, were hurried off to the already overstocked market by producers, intent only on saving what they could out of the wreckage. Although prices sank lower and lower, it was pointed out that there had been no decline in the price of beef to the consumers. Thus the antagonism against the meat packers, which resulted in an investigation of the transportation and sale of meat products by a specially appointed senatorial committee, commonly known as the Vest Committee, had its origin, so far as the range producers were concerned, in the preceding boom years.[53]

The efforts made by stock associations to force the railroads to reduce their rates and improve their service, and the promotion of schemes in the range country to deprive Chicago of its meat-packing supremacy are evidences that the day was over when the pioneer cattleman shipped cheap steers to a market that appeared to him to pay pretty good prices. There were agreements on the part of stockmen to ship only over those lines that were willing to come to terms with their associations on the matter of rates. There were demands, backed by the threat of boycott, that the railroads improve their service. Here was a real grievance, for the

[53] Vest Report, 1888, *op. cit.*

roads had done little or nothing to furnish the proper shipping facilities for the stockman. The cars used in transporting stock were small and ill-adapted to the use to which they
were put. Since the freight charges were based on the car-
load, the tendency was to crowd as many animals as possible
into each car. The poor roadbed and the use of the old
link-and-pin coupling resulted in extreme jolting and jerk-
ing, so that by the time the cattle got to market, there had
been a real depreciation due to the hardships of the journey.
Stock growers' associations insisted on improved rolling
stock and threatened to refuse to use those roads that did not
heed their demands.

The development of the refrigerator car in the seventies
led to much optimistic talk, during the boom period, of
establishing slaughter houses on the High Plains, which
would free the stockman from the commission merchants
and packers. "The day will come," prophesied an enthusiast
speaking before the Wyoming stockmen, "when a live bul-
lock will only be seen in the circuses of Chicago . . . . when
dressed beef will be a home industry and wipe out the commission merchant, but at present they [sic] are a necessary
evil." [54] Optimists, such as Marquis de Mores, envisioned
the range country dotted with packing plants which would
destroy the position that Chicago had attained in the livestock and meat-packing trade. Beyond the operation of
butchering establishments to supply the local trade, this
hope was not realized. First of all, the capital for such
ventures was not available; second, the volume of the range-
cattle business could not be taken up except by large plants
and by a very highly developed system of marketing; and
third, the habit of shipping to a large central market was
strong among the range cattlemen, who looked forward to
the trip eastward on railroad passes with their cattle and

[54] *Cheyenne Daily Sun,* April 8, 1884.

took pride in the beef shipment as it arrived on the hoof in the Chicago stockyards.

Another project, quite as impractical as that of the De Mores packing plants and as characteristic of the boom period, was the scheme to ship cattle for feeders direct from the Montana and Wyoming ranges to England. The exportation of live cattle to England had begun as early as 1868 and by the late seventies had attained considerable proportions. The cattle for this trade had been drawn chiefly from the Middle West, as the English consumer did not take kindly to the far-western range stock.[55] This had forced the English agriculturalist to put up the same losing fight against cheap American beef that he had been waging against cheap American wheat. His feeders, or "store cattle" as they are called in England, came chiefly from Ireland as yearlings, which must be held as feeders for two years before they were ready for market.[56] If cheap American steers, which could be fattened for market in a few months, could reach the unused pastures of England, the advocates of this scheme saw the English stock grower delivered from his plight.[57]

The advantages that such a trade would have for the range man appeared equally great to the proponents of the scheme as they discussed it in the optimistic atmosphere of the Cheyenne Club. Here was a new market which would at once yield a profit and put the Chicago commission merchants and western railroads in a position where they either had to meet the demands of the western range men, or stand

[55] Trimble, 232.

[56] During the ten years, 1875-1884, over six millions of cattle were imported from Ireland, 52 per cent of which were store cattle for fattening purposes. Great Britain, *Parliamentary Papers*, House of Commons, 1884-1885, Vol. 73, Cd. No. 4372, p. 63.

[57] The correspondence between Moreton Frewen and Thomas Sturgis, secretary of the Wyoming Stock Growers' Association, has been used for this episode. This correspondence is found in the files and letter books, 1884-1885, of the Association in Cheyenne, Wyoming.

to lose a part of their western trade. Cattle could be shipped over the Northern Pacific to Duluth and thence over the Canadian lines to Montreal and there loaded on cattle steamers for England. Those who were urging this scheme pointed out that in 1884 bullocks of decent quality brought forty-four shillings a hundredweight in England, which was about nine and one-half cents a pound.. Allowing three cents per pound for transportation charges, this figure meant six cents per pound on the Wyoming range.[58] Enthusiasts saw the entire western range country becoming a breeding place for lean kine which the British farmer could be induced to fatten for market.[59]

There appeared to be but one obstacle to the carrying out of this project, an obstacle, indeed, which might be turned to the western cattleman's advantage. The fear of pleuro-pneumonia, which had developed in the eastern states, had resulted in the English Privy Council acting under a parliamentary statute, issuing an order prohibiting the importation of cattle from the United States except under the most stringent conditions, which included quarantine and slaughtering within ten days at the point of debarka-

[58] Letter, Frewen to Sturgis, Aug. 1, 1885.

[59] The following resolution was passed by the Wyoming Association at its annual spring meeting in 1884.

"*Whereas*: This Association finds it has been seriously injured by the pooling arrangement prevailing among buyers at the Chicago Stock Yards; and

"*Whereas*: The prices of hay and yardage constitute a heavy tax upon the value of cattle: therefore,

"*Resolved*, That it is incumbent upon the Association to examine any and all means which may result in opening another market for our cattle whereby large returns may be obtained: that the plan proposed by Mr. Frewen of shipping beeves via. the Northern Pacific R. R., Duluth and Canada to England offers a possible method of getting our beeves, also lighter cattle if desired to the very highest market, viz: Great Britain; that the Executive Committee is hereby instructed to obtain the fullest information on the above points and report. . . . " *Breeder's Gazette*, V, 544, Apr. 10, 1884.

tion.[60] It was within the discretion of the Council, however, to exempt from the operation of the order, cattle from such portions of heretofore scheduled countries as were free from disease.[61] If the Privy Council could be persuaded into admitting range cattle on the ground that they were free from pleuro-pneumonia, the western cattleman could free himself from the Chicago market and monopolize the cattle export.

Fantastic as this scheme appeared to the more conservative, the fact that it might be used as a club against the railroads and the commission men and as a talking point for promoters of the cattle bonanza, seeking additional British capital, commended it to those who sensed that the cattle boom was about to deflate. The Wyoming Stock Growers' Association delegated the task of persuading the Privy Council to admit range cattle to Moreton Frewen, English manager of the Powder River Cattle Company, who seems to have been the originator of the scheme. This gentleman, whose English connections gave him a ready access to the Government, arrived in London in the summer of 1884.[62] On July 25, 1884, Frewen appeared before Lord Carlingford and Mr. Dodson, members of the Privy Council, with a delegation made up of several prominent Englishmen interested in western cattle raising and including among their number, twelve members of Parliament. Lord Wharncliffe, the chairman of the deputation, who seems to have taken Frewen seriously, introduced him as the representative of the Governor of Wyoming, a territory west of Lake

[60] R. A. Clemen, *The American Livestock and Meat Industry* (New York, 1923), 280-281.

[61] The Law Reports, *The Public General Statutes*, 47 and 48 Victoria, 1884, p. 20.

[62] The *Breeder's Gazette*, in commenting on this incident remarked that, "It strikes us, however, that our Wyoming friends would greatly increase their chances of success by conducting their negotiations through the proper channels at Washington." *Breeders' Gazette*, VI, 254, Aug. 21, 1884.

Michigan. Frewen presented a picture of the millions of lean cattle in Wyoming and Montana which could be fattened by the English farmer, who would be glad to get these cheap feeders. As they were full-grown cattle, the time necessary to mature them for market would be short, as compared with Irish store cattle, which sold in England as yearlings. He pointed out that under Wyoming law, the governor could quarantine against eastern cattle, thereby removing all danger of pleuro-pneumonia getting started among the Wyoming herds. Frewen claimed that he had interviewed the Canadian officials and that both Lord Landsdowne, the governor-general, and Mr. Pope, the minister of agriculture, had shown themselves only too anxious to cooperate.[63]

Neither the British farmers nor the Canadians stood to benefit from such a project. The farmer could hardly be expected to look with favor upon a government policy which would open its doors wide to such a flood of cheap cattle. The market was low enough as it was, without deliberately lowering it still further by encouraging further importation. As to the Canadians, Frewen was wholly without the assurances from the Canadian Government that he had given out in London. The Privy Council would take no action until the Canadian authorities had been consulted. Their reply was unfavorable. They wrote that at the moment, any relaxation of the Canadian quarantine laws which would permit the shipment of Wyoming cattle through Canada was inopportune, even though the northern ranges were virtually isolated from pleuro-pneumonia.[64] The appearance of the

---

[63] The *London Times*, July 26, 1884, gives an account of the deputation and devotes a long editorial to the possibilities of such a trade. A letter from the clerk of the Privy Council to the Colonial Office, dated Aug. 5, 1884, gives a full account of the proceedings. Great Britain, *Parliamentary Papers*, House of Commons, 1884-1885, Vol. 73, Accounts and Papers 19.

[64] *Op. cit.*, 9-10.

Texas fever on the Montana and Wyoming ranges in 1884-1885 was excuse enough for the Government to veto a scheme, which, if put into effect, would crowd the Canadian stock grower.[65]

By 1885 two things are clear to the student of the cattle range industry; first, the business was ceasing to be a frontier industry, and second, it was falling a victim to over-expansion. It was true that the frontier environment was still there. There were vast areas of land, unoccupied save for herds of cattle and scattered ranch houses. The danger, the hardship, and the isolation, which the frontier imposes, were still present. The Indian still rode down from the reservation to steal cattle and murder lonely cowboys. Cattle thieves were still the scourge of the ranges and required the frontier remedy of violent and sudden death. But all these conditions were being borne to an ever increasing degree by cattlemen for the benefit and profit, not so much of themselves, as of a company with offices in Boston, New York, or Edinburgh. The early herds, which had come in on the ranges in charge of their owners, were disappearing, as outfit after outfit became part of a company, capitalized in the East. Although the small operator did not altogether disappear, his influence in the cattle community became less and less as compared with that exercised by the owners or managers of larger companies. It was becoming more and more difficult for the man of small means to stay in the business. Somehow or other, his cattle disappeared, or if he succeeded in maintaining the size of his herd, he became an object of suspicion as a rustler. Thus the opportunity which the frontier had always extended to individual enterprise, and which is one of the evidences of its existence, was fast disappearing.

[65] Letter, Frewen to Sturgis, undated, Wyoming Stock Growers' Association, 1885 file; *Breeder's Gazette*, VI, 420, Sept. 18, 1884.

Finally, in the scramble for profits that had resulted from inflation and speculation, the business had extended far beyond the margin of safety. Cattle were pushed into areas where the grazing was poor or crowded into already overstocked pastures. Then, and not for the last time, the semiarid West witnessed a retreat of those who asked too much from an environment demanding an organization and a technique which they as a group had not yet acquired.

V

ORGANIZATION

THE period from 1879 to 1885 was marked by an enormous expansion of the range business, by a sharp increase in the amount of capital invested, and by a crowding of the ranges. The questions arising over the ownership of cattle and the rights of grazing, difficulties that have bothered the pastoral industry from the beginning of time, were intensified as the number and value of the herds increased. Millions of head of cattle, representing very nearly the total capital investment of the range industry, wandered unprotected over vast areas of unoccupied land. The task of affording protection to this scattered and exposed property was too great for either the local or the central governments. The physical basis of the industry was the public domain, where one man's right to free grass was as good as another's. To expect the Federal Government to pass legislation that would assure to each grazier his share of the grass was to call for too wide a departure from the ideas upon which our public land policy was founded. Thus, in the absence of government protection and support, the range cattlemen had to depend upon their own ability to create organizations which would be strong enough to enforce the customs indigenous to the range country, which would strengthen the local governments in the enforcement of law, and which would meet the new problems that arose as the industry grew in size and importance.

In a study of the development of cattlemen's organizations, one must keep in mind the fact that on this last frontier, social groups were slow in forming and extremely loose in organization. The advantages of isolation were obvious. Since the cattleman could not afford to buy the land upon which his cattle grazed, and since he was not allowed to lease the public domain, his ideal was to find a large, isolated area with as few neighbors as possible. Little cooperative effort would be needed under such ideal conditions to produce or to get his product to market. He was perforce unsocial. He did not seek cooperation, but came to it when the presence of neighboring herds forced it upon him. Unlike the farmer, his financial rewards were greatest when his isolation was the most complete. Cooperation did not come in order to remedy the evils of isolation, but to meet the problems created by the presence of too close neighbors. The frontier farmer cooperated because such action meant roads, rising land values, and social amenities; the cattleman cooperated to preserve as best he could the conditions that were naturally his through isolation.

"The permanence of a common aim and will ruling over the decisions and interests of single members has been declared to be the test of the corporate character of associations," writes a distinguished student of social origins.[1] By such a test, the common aims of the cattle community were threefold: first, to preserve the individual's ownership in his herd and its increase; second, to afford protection to the individual's herd; and third, to control the grazing of the public domain in order to prevent overcrowding. These aims, which might have been achieved by an individual in the earlier days of comparative isolation, could now only be realized through group effort.

The preservation of the individual's ownership in his herd and its increase led to regulations concerning marks and brands, the roundup, the maverick, and the control of bulls on the range. Some of these regulations were a part of our

[1] P. G. Vinagradoff, *The Growth of the Manor* (London, 1905), 139.

legal codes and enforceable by the civil authority, some became part of the codes of the western states "by that process by which in a new land social customs form and crystallize into law";[2] and some formed an extra-legal code, its sanction depending solely on the group will.

The cattleman found it necessary to protect his property from Indians, thieves, wild animals, and disease. This task belonged properly to the government, local and federal. The story of the American frontier, however, is full of examples of local governments too weak or too disorganized or too corrupt to afford protection and of a federal government, whose power became feeble when operating through subordinates in a sparsely settled region far from the seat of government. When such conditions obtained, the frontiersman, whether he dwelt upon the upper Tennessee or the Rio Grande, whether he hanged road agents on the American Fork or at Virginia City, or galloped after horse thieves, white or red, over the sage-brush flats of eastern Montana or Wyoming, displayed the same characteristics. He joined with his fellows for mutual protection.

The problem of controlling the grazing on the public domain in order to prevent overcrowding and to preserve the individual's share, was the most difficult that the group had to face. As an individual American citizen, the cattleman might use his rights under the land laws to obtain his share of a small portion of the public domain.[3] He might get such an allotment along the water front, thereby exercising con-

---

[2] F. J. Turner, "Middle Western Pioneer Democracy," *The Frontier in American History* (N. Y. 1920), 344.

[3] Under the Homestead Act, the Timber Culture Act, and the Desert Land Act, a settler might obtain 960 acres of land: 160 by residing thereon for five years and cultivating the same, 160 more by planting ten acres of trees (nuts, slips, or cuttings), and 640 more by paying down twenty-five cents an acre, irrigating, and by paying a dollar an acre more at the time of final entry.

trol over grazing, which was valueless to anyone who did not have access to the water. In no other way could he legally control any of the public domain. Beyond that, his control must be extra-legal and at times illegal. He might keep off other herds either by fencing the public land or by cooperation with his neighbors. His success in these methods of trying to solve the problem of maintaining his share of the free pasturage depended, on the one hand, on the negligence or impotence of the central government in the enforcement of the land laws, and on the other, on the sanctions of range customs backed, if necessary, by group force.

As an individual, he might bargain with the railroads for the purchase or lease of railroad sections, or with the state or territorial governments for some of their lands. Deals might be made between him and some Indian agent for privileges of grazing on a reservation, transactions sometimes frowned upon by an Indian Bureau that for years did not know its own mind on the matter. Through his organizations, he could give weight to the constant demands for the reduction of the reservations, thus extending the grazing areas. In a group, he could protect the ranges from grass fires by punishing those who were guilty of setting them and, by working with his neighbors, he could help to put out such fires before they spread and destroyed a whole season's grass.

In a study of the rise and development of the cattlemen's organizations, formed as a result of these conditions, one can watch the characteristic frontier individualism succumb to the equally characteristic frontier need for group effort, the evolution of custom into law, and the appearance of certain institutions, which became part of the economic and social structure of the Far West.

As long as cattle on unfenced common are valuable enough to pay for the trouble of identification as individual property, groups of owners will cooperate naturally in the task of herding, gathering, branding, and protecting their property. Although the price of cattle was very low in Texas in the sixties before the northern market was opened, they were gathered and branded by the owners on a "cow hunt," the forerunner of the roundups of the range country. The early drives to the railroads were usually cooperative affairs, for the beef herd might contain any number of different brands. This fact meant that all or a part of the owners of a given district had an interest in the trail herd. These brands were listed before the drive began, and, on the return of the drovers, a settlement with the owners was made at a meeting called for that purpose.[4] As soon as it was demonstrated that money could be made out of driving cattle to the railroads and shipping them eastward, the cattle thief arrived in Texas. As early as 1868, two years after the first drive, small groups of owners were organizing themselves into protective associations and hiring stock detectives.[5] Thus, at the very outset, there appeared those institutions and figures that were a part of the range-cattle business wherever it was carried on: the roundup, the association, the trail-driver, the stock detective, and the cattle thief.

As the range industry spread northward, stock growers' associations sprang up. Colorado stockmen felt the need for common action when the Texas herds began to come into their territory. Trail herds, headed for the northern ranges, passed through the Colorado herds, and the drovers were far from careful to see to it that none of the local cattle were picked up as the drive moved along. By 1872, two stock associations were in existence in Colorado; the Colorado

---

[4] *Supra*, pp. 32-33.

[5] The first formal association for protective purposes in Texas which the author could find was organized in 1868 among the stockmen of Limestone, McLennan, Falls, Hill, and Navarro counties. J. F. Dobie, "Detectives of the Cattle Range," *Country Gentleman* (Philadelphia), February, 1927, p. 30.

Stock Growers' Association, organized January 19, 1872, and the Southern Colorado Association.[6] At the meeting of the former, in 1873, stock growers from the adjacent territories were invited to attend, for it was found at the outset that the new industry was not limited by state and territorial lines. At this meeting, plans for the conduct of roundups, for the suppression of theft, and for the regulation of brands were discussed.[7] The close connection between the interests of the Colorado stock growers and those of Wyoming was evidenced by the fact that sixteen delegates from the Wyoming Association were present at the meeting.[8]

The first organization of Wyoming stockmen occurred a few months prior to that of Colorado. On October 28, 1871, an organization meeting was held at Cheyenne, and in the following month, November 14, the first annual meeting of the Wyoming Stock Graziers' Association was held. This first body was composed chiefly of local residents of Cheyenne, many of whom were engaged in the freighting and stage business and felt the need of some action to protect their property. The necessity for more stringent laws against stock theft was felt, and the question of getting the next territorial legislature to pass such legislation was discussed. Beyond this, the activities of this first association did not go, nor does it appear that its life extended beyond this one meeting.[9]

Down along the North and South Platte and their tributaries, where the Texas herds were arriving each year in

[6] *Third Annual Report of the Denver Board of Trade* (Denver, 1871), p. 61. *Cheyenne Daily Leader,* Jan. 13, Jan. 20, Jan. 26, 1872.

[7] *Cheyenne Daily Leader*, January 21, 1873.

[8] *Ibid.,* January 6, 1873.

[9] *Cheyenne Daily Leader,* October 30, 1871. Among those who were interested in this earliest attempt to form an association were the first territorial governor, J. M. Campbell, Dr. Latham, J. M. Carey, W. W. Corlett, later attorney for the Association, and S. F. Nuckolls.

increasing numbers, those who were actually engaged in range herding began to find that organization was necessary, not only for protection but also for the conduct of their business, which had now reached the point where it could not be carried on without mutual assistance. Here, among these early range cattlemen, there was created the organization that later developed into the most powerful of all the associations of the plains. In giving the reasons for the formation of the Laramie County Stock Growers' Association, one of the original members writes:

The growth of the cattle business up to and including 1873 had been rapid; many had engaged in it. The country where most of the stock had been turned loose was east of the mountains and between the Platte rivers as far east as Ogallala, Nebraska; the cow camps growing closer each year. When a stock owner wished to work his cattle, he would send word to his neighbors and all would round up, get their stock, brand calves, turn loose and drive home. But so many outfits had come in and rounded up the stock, and ginned them over so much, they could never get fat. This continual working over and over of cattle was detrimental to the business, and those interested . . . wanted some plan or system laid down.[10]

Thus frontier individualism surrendered to economic necessity, and on November 29, 1873, the Laramie County Stock Growers' Association was organized. On May 15 of the next year, the first organized roundup on the Wyoming ranges was held under the direction of a foreman appointed by the Association and in accord with a plan of work laid down by that body.[11] From then on, the expansion of the range-cattle industry in Wyoming can be measured by the growth of the Association and the extension of its influence. Eleven men were present at its first meeting. At its second

[10] Letter of T. H. Durbin in *Letters of Old Friends and Members,* 47.

[11] Minute Book of the Laramie County Stock Growers' Association in the possession of the Wyoming Stock Growers' Association, p. 3; also *Cheyenne Daily Leader,* May 2, 1874.

meeting in the spring of 1874, this number was increased to twenty-five.[12] In 1876 two roundup districts were laid out, and in 1878 four were provided for.[13] In 1879 the name was changed to the Wyoming Stock Growers' Association.[14] At its meeting in 1881, it was decided to invite the local stock associations in other parts of Wyoming and the associations of Wild County, Colorado; Rapid City and Deadwood, Dakota; and Lincoln, Keith, Cheyenne, and Sioux Counties, Nebraska, to send delegates empowered to effect a consolidation.[15] In response to this invitation, the association of Sioux County, Nebraska, and Albany County, Wyoming, joined in 1881.[16] In the next year, the stock growers of Uinta, Carbon, and Johnson Counties joined the central association.[17] By 1883, the association had 267 members on its rolls, a membership which, two years later, had increased to 363, owning some two million head of stock.[18]

In Montana, an attempt at organization was made at about the same time. In December, 1873, a group of stockmen met in Virginia City and issued a call to the stock growers and ranchmen of Madison County to meet the following month. It was announced that the purpose of such a meeting was to consider and discuss

. . . . the best method of protecting the winter ranges from summer grazing; the disposition of estrays, the rules and regulations that should be adopted in relation to brands; for the better protection against cattle thieves; for the purpose of securing general unanimity of action among the stockmen of the County; for the purpose of taking necessary steps towards the organization of a stock growers' association and to take measures to secure the required legislation at the

[12] Minute Book, 4.
[13] Ibid., 7, 11-12, 22.
[14] Ibid., 32.
[15] Ibid., 57.
[16] Ibid., 64.
[17] Ibid., 74-75.
[18] Ibid., 86; Clay, 250-251.

forthcoming (Territorial) Assembly. . . . That action among stockmen is required, is apparent to all.[19]

On January 8, 1874, the meeting was held, a chairman was
elected, and a committee appointed to draft a bill for the
regulation of the cattle industry. This bill was presented
to the legislature then in session. No permanent organization was perfected, and the task of achieving the objects for
which the meeting was called was left to the law-making
body of the Territory.[20]

No further evidence of common action on the part of the
cattlemen of the Territory appears until 1879. By that time,
it had been clearly demonstrated that the laws which had
been passed by the territorial legislature had failed to regulate and protect the industry. If the stockmen were to get
what they wanted, it would have to come about through their
own efforts. Consequently on January 23, 1879, a group,
fairly representative of the industry, met in Helena. James
Fergus, one of the oldest and most influential stockmen in
the Territory, was present and urged the formation of a
stock growers' association. A committee was appointed to
consider the propriety of forming such an association, and
to draft a constitution and a set of by-laws. Such prominent
stockmen as Conrad Kohrs, Ancenny, and Poindexter were
on this committee. Sutherlin, editor of the *Rocky Mountain
Husbandman*, was also a member.[21]

On February 19, 1879, the first meeting of the Montana
Stock Growers' Association was held at Helena.[22] A constitution was adopted and officers elected. There were twenty-
two signers to the constitution of the new association. The
president, R. S. Ford, a Sun River stockman, issued a call

---

[19] The *Weekly Montanian*, December 18, 1873.
[20] *Ibid.*, January 15, 1874.
[21] *Helena Independent*, January 25, 1879.
[22] *Ibid.*, February 25, 1879.

through the papers to all Montana stock growers, urging them to form themselves into district or local organizations and unite with the general organization at its next meeting the following September. Local associations were organized or reorganized in Madison County, the Sun River country, Lewis and Clark County, the Hamilton district in Gallatin County, Deer Lodge County, the Shonkin district in Choteau County, and the Smith River district in Meagher County. The areas represented by these local organizations constitute that section of the Territory where the industry had developed chiefly as a result of the growth and expansion of the native herds.

After a third meeting of this organization, held March 3, 1880, there does not appear to have been any further general meeting in western Montana until another reorganization occurred in Helena, July 28, 1884.[23] Through his local organizations, the cattleman had, in the interval, developed a workable roundup system and had, in some cases, notably in the central and eastern sections, taken vigorous action to protect his property from Indians, thieves, and fire.

The meeting held in 1884 marks the beginning of the Montana Association's uninterrupted career. From then on meetings were held semiannually. At this meeting, eleven of the fourteen counties of the Territory were represented. Delegates were elected to attend the meeting of the Eastern Montana Stock Growers' Association in Miles City. This latter association had been organized in 1883, when the sudden expansion of the range-cattle business in eastern Montana over the former hunting grounds of the northern tribes, had resulted in Indian raids that threatened to exterminate

[23] *Rocky Mountain Husbandman*, August 7, 1884. Granville Stuart in his *Forty Years*, II, 171, speaks of a meeting held on August 15-16, 1882, at Helena, where 168 were in attendance. No mention of such a gathering could be found in the territorial press of that date.

the herds north of the Yellowstone.[24] At its third meeting on April 3, 1885, held at Miles City, members from the Montana Stock Growers' Association were in attendance, and a combination was effected that united the newer range interests of the plains with the older interests of the mountainous western section of the Territory.[25] At the time of this consolidation, it was decided to hold the spring meeting in Miles City and the fall meeting in Helena.

More important than the brief account of the rise of the cattlemen's organizations, sketched above, is a study of their efforts to meet and solve the problems that were a part of this business of raising cattle on the open range, and of the effects that such attempts had on the political and social life of the community. Whenever his isolation was destroyed, the range cattleman, in spite of his apparent independence, needed a degree of community protection and encouragement, which neither the farmer nor the miner demanded. In some cases, these needs were satisfied under the law that afforded protection to all, and, where such protection was inadequate, special legislation was resorted to. There were, however, certain difficulties inherent in the business that could not be met within the law at all; these were left to extra-legal and sometimes illegal action.

In regions where the cattle-growing interest was paramount, the activities of the stock growers' association, legal or extra-legal, as the case might be, received general support; for the welfare of the whole community rested to a very large degree on their prosperity. In those sections where other interests had developed, such as mining or agriculture, the case was not so simple. There interests clashed, and the cattleman could not hope to exercise the

---

[24] Minute Book of the Montana Stock Growers' Association, MS in the possession of the Montana Historical Society at Helena. See also the *Weekly Avant Courier* (Bozeman, M. T.), April 26, 1883.

[25] Minute Book, 1-51.

control over the affairs of the community that the nature of his industry appeared to require.

A comparison of the activities of the Wyoming Stock Growers' Association with those of the Montana Association will serve to illustrate this point. For at least a decade, the former was the unchallenged sovereign of the Territory of Wyoming; the latter never attained such a position of dominance, because a combination of the mining and agricultural groups in the western section of the Territory was more than sufficient to outweigh the stock-growing interest even in its best days.

When cattle are turned out on unfenced public domain, the necessity for some means of identification is obvious. The practice of branding and marking cattle was not peculiar to the unfenced High Plains, but had been used by stock owners on every frontier where stock had been grazed on commons. The enactments of the early territorial legislatures of Wyoming and Montana concerning marks and brands could be duplicated in the codes of the older states. Both legislatures, in their first sessions, passed laws requiring each stock owner to adopt a distinctive brand, which must be recorded with the county clerk, who must not record the same brand for more than one resident of the county. Punishment by fine or imprisonment was provided for those who knowingly and willingly used a brand already recorded or who attempted to deface or destroy a brand.[26] Both legislatures passed the usual law covering estrays, containing provisions for the taking up of stray stock, the advertising of them for a certain period, and their sale by the county, if they went unclaimed.[27]

In a country of limited ranges and small herds, the legal

[26] *Laws of Wyoming Territory*, 1869, Sess. 1, pp. 426-427; *Laws of Montana Territory*, 1864-1865, Sess. 1, p. 401.

[27] *Laws of Wyoming Territory*, 1869, Sess. 1, pp. 708-711; *Laws of Montana Territory*, 1864-1865, Sess. 1, p. 404.

protection outlined above would have been sufficient. Wherever and whenever the range-cattle industry developed, such laws were found to be wholly inadequate. In Wyoming, the arrival of the Texas herds in the seventies resulted in each legislature passing laws to adjust the brand system to the changing character of the business. The drover who brought cattle to or through the Territory must see to it that every head in his herd was branded. He must frequently examine his herd and drive away any cattle not his own. Because whole brands of cattle were changing hands, provision was made for the lawful purchase of a brand. Penalties were provided for those who failed to brand any animal over a year old, who used a "running brand," who failed to obtain a bill of sale with a full list of the brands of the animals purchased, who killed an unbranded calf, or who skinned an animal carrying another's brand, unless he could produce evidence of purchase.[28] Conflicts over brands, which had been left to the county clerk for decision, were, in 1877, turned over to a committee composed of the clerk and two resident stock growers of the county; for with the increase of herds, the brand system became so intricate that it required the knowledge of the community to administer it. All owners bringing cattle into the Territory, were required to lay the brands of these cattle before the committee, which was instructed to reject all brands that were duplicates of existing brands. The addition of a circle or a half circle, a bar or a box, did not create a new brand and must be rejected.[29] In 1879, all drovers were required to brand with a road-brand before driving over any portion of the Territory. Such a brand would distinctly set off trail cattle from all others. At the same time the law on illegal branding was strength-

[28] *Laws of Wyoming Territory*, 1871, Sess. 2, p. 90; 1873, Sess. 3, p. 226; 1875, Sess. 4, p. 242.
[29] *Ibid.*, 1877, Sess. 5, p. 124.

ened by making such an offense a felony with a penitentiary term attached.[30] Similar legislation in Montana as to the recording of brands, the changing of brands, and the driving off of stock was passed at about the same time.[31] Not until 1881, however, when the arrival of thousands of Texas stock in eastern Montana made it imperative, did the Montana legislature pass a road-brand law.[32]

On a sparsely settled agricultural frontier, where the farmer turned loose his small band of cattle to graze on the unused public land, the task of looking out for his stock was not a serious one. When they strayed, the farmer and his boys turned out and drove them back home. If a head or two were missing, he watched the estray advertisements published by the county clerk. A law against stock stealing, properly enforced, and an adequate estray law gave him all the legal protection he needed. For the range cattleman, this was not sufficient. His stock, together with thousands of head belonging to other owners, were scattered over hundreds of square miles of territory. Although a range steer might drift a hundred miles from home, it was in no sense a stray, nor was its owner desirous of having it held for advertisement and sale as an estray. Furthermore, the job of getting his property together when he did want it was more than the individual owner could easily perform.

In Wyoming, where the range industry was carried on to the exclusion of almost all farming operations, the estray law came to be practically a dead letter. Here, when the increase in the number of cattle made cooperation economically essential, the roundup system developed outside of the legal framework. Beginning with the Laramie County roundup in 1874, the Wyoming stock growers through their

[30] *Laws of Wyoming Territory*, 1879, Sess. 6, p. 132; p. 135.
[31] *Laws of Montana Territory*, 1869, Sess. 5, p. 100; 1871, Sess. 6, p. 287.
[32] *Ibid.*, 1881, Sess. 12, p. 59.

association, ordered, laid out, and regulated this activity with no power to enforce their regulations, save the negative one of refusing to cooperate with, or afford protection to, those who would not obey the will of the group. Not until 1884, when the roundup system was perfected in Wyoming, was it recognized by the law and punishment provided for those who attempted independent roundups.[33]

In Montana, the case was somewhat different. In the western section of the Territory, cattle growing on a large scale was being carried on during the seventies. Agriculture was also developing in the fertile valleys, as the mining population grew and the local market expanded. Because the expansion of the cattle industry was limited during this decade by the Indian barriers to the north and east, the large operator found himself restricted to those same areas where farming and stock growing on a small scale were arising. It was to be expected that the Montana cattleman trying to operate on a range basis would find plenty of difficulties in adjusting his business to that of the small operators around him. His cattle wandered across county lines or onto unfamiliar ranges and were seized by county officers, who sold them unless they were claimed within a certain period. When he rode out to gather up his stock, he found the farmers of the neighborhood disinclined to drop their work in order to spend several days rounding up, an operation that appeared to benefit the larger operator only; for the farmer's stock stayed near home where there was a haystack. When the large operator bought a few head from the farmer, he found that if the newly purchased animals wandered back home and mingled with the old herd, he might have some difficulty in proving ownership, even though he had placed upon them his own mark and brand.

[33] *Laws of Wyoming Territory,* 1884, Sess. 8, pp. 148-152.

In 1874, a law was passed creating a roundup system, which it was hoped would force all operators, large and small, who turned stock out on the common, to cooperate. The statute required that the county commissioners should divide the respective counties into stock districts of suitable size, and that twice a year they should order the stock in these districts to be rounded up under such rules as they should lay down. These roundups, however, were not to include work cattle, dairy or milch cows, or cattle under the supervision of a herder; in other words, all cattle were exempt save those grazing on the open range. The stock growers in each district were required to meet and choose a stock board. This board was to adopt a district brand and vent, and all estrays were to be branded with the district brand and, if sold or returned to the owner, vented with the district vent. If estrays were not claimed after two roundups, they were to be sold to the highest bidder, the clerk of each board keeping a record of all unclaimed animals. Anyone selling cattle must vent at the time of sale, and failure to do so was punishable by fine.[34]

This attempt to create a roundup system by law was a failure, for the reasons that have been suggested above. Although roundup districts were named by the county commissioners in each of the organized counties of the Territory, only the larger owners appeared on the date set.[35] The estray provision in the new law was not a sufficient modification of the old estray law to satisfy the larger owner, who was beginning to get the range idea that if a cow wandered to the ends of the earth, his brand on her flanks

[34] *Laws of Montana Territory,* 1874, Sess. 8, pp. 85-90.

[35] A note in the *New Northwest,* May 16, 1874, states that, "The first of the week they attempted a 'whoop-up' [*sic*] in Capt. Williams district on the Stinking Water, but it was a miserable failure." The term "roundup" was not in common use in Montana at this time.

gave him ample protection whether he claimed her or not. In time she would drift back to the home range.[36] At the next session, the legislature eliminated the mandatory feature of the law and provided that stock districts might be organized if the local stock growers so desired.[37] The next legislature repealed the law altogether.[38] In the House Committee report on the repeal, it was pointed out that since the law bore hard on the small rancher, who was carrying on agriculture in connection with his stock growing, and who could not attend the legal roundup where his cattle might go unclaimed and be sold, and since it was impossible to pass a law that would satisfy such divergent interests, ". . . . the stock growers of each county or district should be allowed to regulate their own affairs or form stock growers' associations and be governed thereby. Roundups for stock are actually necessary in some localities, but they should be governed by the stock growers themselves."[39]

Thus it was left to the Montana cattlemen to follow the example set by those of Wyoming and develop their own system. "We want to establish a system," declared President Ford at the meeting of the Montana Stock Growers' Association in Helena, 1880,

. . . . that allows each individual stock grower to retain all the rights and privileges he now enjoys, and add to those privileges a system that will not only compel but encourage and even pay men to be

---

[36] The courts at a later time defined the estray so as to meet the range conditions. In an Oregon case, the court stated that under range conditions, an animal was expected to roam until it was gathered up at a roundup. It became an estray only when it wandered into a distant locality and became lost to its owner. 8 *American State Reports* 270, Stewart *v.* Hunter, February, 1888. See also *Rocky Mountain Husbandman*, April 3, 1879.

[37] *Laws of Montana Territory*, 1876, Sess. 9, pp. 109-111.

[38] *Ibid.*, 1877, Sess. 10, p. 237.

[39] Report of the committee quoted in the *Helena Daily Herald*, February 3, 1877. The committee found it necessary to put in quotation marks and to define the term *maverick*.

honest; a system that protects our stock on any part of the Public
Domain the same as on our home range; a system that will put a
stop to animals being taken off the public range and advertised as
estrays; a system that will not condemn our stock as estrays and offer
them for sale, because they were driven or strayed from their home
range; a system that protects in proportion to the stock owned and
one that we all pay for in proportion to our protection; a system that
will enable us to know the marks and brands and vent brands of
every stock grower in the Territory; a system that is more than
democratic in its broadest sense, not only doing the greatest good
to the greatest number, but one that will result in great benefits to
every stock grower; a system that will prevent our territory from
becoming a paradise of thieves; a system that resolves stock growers
into a protective force and pays them to look after the interests of
one another, inasmuch as our interests are identical.[40]

Although the general association did but little to meet these
requirements laid down by its president, strong local associations in the central and eastern sections of the Territory
were organizing and regulating the industry, which, in those
regions, was exclusively a range-cattle business.

As has been stated, the chief reason for cooperation in
rounding up cattle on the range was to prevent the over-
working, caused by the successive roundups of different
owners.[41] At first, each district managed its own roundup,
setting the time, laying out the plan of work, and regulating
the conduct of the participants.[42] Little attention was paid to

[40] *Rocky Mountain Husbandman*, March 11, 1880.

[41] *Supra*, p. 120.

[42] The captain in charge of the Shonkin Roundup was given full power
to discharge any member whether cattle owner or hired man who had failed
properly to perform his duty. He was instructed to see to it that horses were
not abused or over-ridden. The bonds of love and understanding that were
supposed to exist between the cowboy and his steed were not quite as
apparent to stock growers as to later fictionists, and the owners in their
local and territorial associations passed very stringent regulations to stop
such abuse of property by their hired men. Card playing, gambling, and the
bringing in of whiskey were rigidly prohibited. Forty dollars a month was

the doings of cattlemen in adjacent districts. As the herds
increased and mingled with those of neighboring districts,
there arose the necessity for common action throughout the
whole range. In some cases, this led no further than agreements to hold the roundups on or near the same date. In
others, however, a consolidation of district associations into
a general association was achieved. By the latter process,
the Laramie County Association in Wyoming, which had at
first been limited to the southeastern section of that Territory, expanded until, as the Wyoming Stock Growers' Association, its jurisdiction was extended not only over all the
Territory but into the neighboring states and territories.
Representatives from Nebraska, western Dakota, and eastern Montana, as well as from every county in Wyoming, sat
upon its executive committee, so that concerted action might
be brought about.[43] The Montana Association never controlled the roundup system for the whole Territory, for
during the whole range period, the local associations in the

agreed to as the amount the members would pay the men on the roundup.
Rocky Mountain Husbandman, May 13, 1884. The number of men and
horses that each member furnished depended on the number of calves
branded by each outfit in the previous year. The Shonkin Association required three horses and one man for every two hundred calves. *River Press*
(Fort Benton), December 28, 1881.

[43] The composition of the executive committee of the Wyoming Stock
Growers' Association was as follows:

Laramie County, Wyo.............3			Crook County, Wyo..............1		
Albany " " 3			Cheyenne " Neb.............2		
Carbon " " 2			Sioux " " 2		
Unita " " 2			Southern Nebraska 1		
Johnson " " 2			Eastern " 1		
Sweetwater " " 1			Dakota 1		
Fremont " " 1			Montana 1		

Total 23

By-Laws and Report of the Wyoming Stock Growers' Association, 1884
(Cheyenne, 1884), 2-3.

central section, the Musselshell, Smith River, and Shonkin roundups, continued to regulate their own affairs. In eastern Montana and the tributary Little Missouri country in Dakota, sixteen roundup districts were laid out by the Association.[44]

In spite of all the care taken by the associations to get the members to agree upon a certain date to start the round-ups, there were some who insisted on staging independent roundups before the date set. As there was no law to pre-vent this, concerted action rested wholly on the boycott that the other members might place upon him who violated the agreement. Even the powerful Wyoming Association was powerless to prevent it. Resolutions were passed binding all members to wait for the date.[45] But the temptation of the spring range with its thousands of unbranded calves was too great, and each year "sooners" were out. Not only was there a loss in weight as a result of these independent roundups where cattle were worked over a number of times, but there was also an actual reduction of the numbers of head on the range. If the cattle were worked too early in the spring before they had shed the long hair that had protected them from the rigors of the winter on the range, they were easily overheated and, when once wet, dried off slowly. When the cold, damp winds of the early spring hit them in this condi-tion, the weaker ones died.[46] The law of 1884, which the

[44] Minute Book of the Montana Stock Growers' Association, 8.

[45] Minute Book of the Wyoming Stock Growers' Association, 63, 68, 71. Some of the worst offenders were contractors supplying beef to the Indian agencies. They often kept part of the herd contracted for by the Government out on the range. As they were needed at the agency, the owner would go out on the winter range and bring them in. The heavy coats of the cattle made the brands almost indistinguishable, which resulted in a loss to nearby cattlemen. Report of the Sioux County (Nebraska) members of the executive committee to the secretary of Wyoming Stock Growers' Association, in Cor-respondence, W. S. G. A., April 31, 1882, File, January-June, 1882.

[46] *Cheyenne Daily Sun*, January 30, 1884.

Wyoming cattlemen put on the statute books, put an end to this evil so far as that Territory was concerned.[47] From then on, the date set by the Association was the legal date, and any one attempting to beat his neighbors out onto the ranges was liable for punishment. By means of an efficient corps of inspectors, stock detectives, and attorneys this law was enforced during the range period. The Montana Association failed to get such legal sanction and was forced to depend upon the boycott to punish recalcitrant members.

The maverick, that by-product of the western range-cattle industry, was a constant source of difficulty. The decision as to the ownership of the increase in the herd rested wholly on the fact that a calf will follow its mother and, no matter how large the herd may be, a cow will know her own calf. In the spring roundups, then, the calves were given the brand of the cow that they were following, such being all the proof necessary to establish the question of ownership. If, however, the mother had died or the two had been forcibly kept apart, then the ownership of the animal could not be established. In the early days of the range, when herds were not grazing over the same area, it was fair enough to assume that the mavericks found in the spring upon a certain range were the property of the man whose herd was accustomed to graze thereon. Thus the doctrine of the "accustomed range" was adopted as a fair method of settling ownership. As long as there were no other herds in the neighborood, such a system would operate. When other herds arrived, the owners could hardly be expected to surrender whatever claims they might have to mavericks simply on the basis of prior occupancy. In some sections, the practice of assuming that the mavericks were the property of the largest female herd in the vicinity was tried, but here again,

[47] *Laws of Wyoming Territory,* 1884, Sess. 8, pp. 148-152.

other owners were denied possession of what might be theirs.[48]

Thus there was an additional inducement to get out on the range ahead of one's neighbors. Once a brand was on a calf, it was hard to prove in court that it had been following a cow of a different brand, which was claimed to be its mother. Furthermore, it was possible to make mavericks by separating some of the calves from the herd and holding them for a time until they could no longer find and follow their mothers. This was a long process and might be detected. "Maverickers," as they were known on the western ranges, found a quicker and more effectual scheme. By slitting the tongue of a calf, it was no longer able to suckle and would soon stop following its mother.

The obvious advantage that the early independent roundup had in the matter of gathering up the mavericks and branding them and in "mavericking" a few calves from the neighbors was another reason for holding all stock owners to a certain roundup date and insisting on concerted action in the roundup. If this were done, then each outfit had as good a chance at the mavericks as any other. In Wyoming, however, the stock growers went still further and made the orphan calf the property of the Territory with the Stock Growers' Association as trustee. In the law of 1884, the maverick was defined and directions carefully laid down as to his disposal.[49] The foreman of every roundup district was instructed by the law to take up all mavericks found by him or his subordinates and offer them for sale at auction on the range every ten days while the roundup was in progess. Upon their sale to the highest bidder, the foreman was instructed to brand each animal with the brand of the pur-

<hr />

[48] Annual Report of Secretary Sturgis of the Wyoming Stock Growers' Association for 1883, Minute Book, 88-89.

[49] *Laws of Wyoming Territory*, 1884, Sess. 8, pp. 148-152.

chaser and with that of the Association. Careful records of such transactions were to be forwarded to the secretary of the Association and kept open for public inspection. The money from the sale of all mavericks was to go into the treasury of the Association to be used in the payment of inspectors and for other like purposes. The foreman was enjoined by the law "to give especial heed to the care of all branded neat cattle, the ownership of which is unknown" and was given full authority to brand the calves of such cattle with the unknown brand. Any member of the roundup disobeying the orders of the foreman in this particular was to be punished by fine or imprisonment or both. In addition to this, the brand law of the Territory was amended by prohibiting the branding of any stock between February 15 and the date set for the spring roundup by the Association. Any violation of this provision was punishable by fine or imprisonment.[50]

The law of 1884 was of very great importance in the history of Wyoming. It made the Association a quasi-public institution with full legal control over the stock industry of the Territory and with power to enforce its regulations as to roundups and brands. Since the Association had full control over the admission of new members, it was possible, by excluding the recalcitrant ones, to bring them to terms, for it would be next to impossible to operate on a range cattle basis outside of the Association. True, the law provided that any person who was directly interested in stock growing could become a full member upon application and compliance with the by-laws, provided he had not been convicted of a felony nor possessed a reputation notoriously bad. This last proviso was capable of considerable interpretation, so that the law did not remove the right of the Association to determine its membership and to a very large degree

[50] *Laws of Wyoming Territory*, 1884, Sess. 8, p. 153.

THE PRATT & FERRIS CATTLE COMPANY.

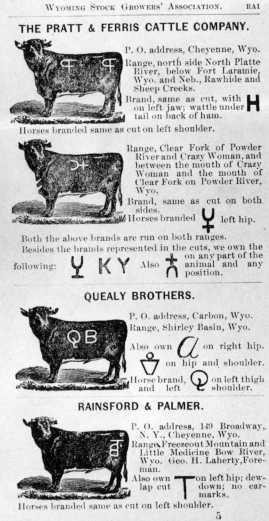

P. O. address, Cheyenne, Wyo.

Range, north side North Platte River, below Fort Laramie, Wyo. and Neb., Rawhide and Sheep Creeks.

Brand, same as cut, with **H** on left jaw; wattle under tail on back of ham.

Horses branded same as cut on left shoulder.

Range, Clear Fork of Powder River and Crazy Woman, and between the mouth of Crazy Woman and the mouth of Clear Fork on Powder River, Wyo.

Brand, same as cut on both sides.

Horses branded ⚲ left hip.

Both the above brands are run on both ranges.

Besides the brands represented in the cuts, we own the

following: ⚲ K Y   Also 𝄞 on any part of the animal and any position.

QUEALY BROTHERS.

P. O. address, Carbon, Wyo.

Range, Shirley Basin, Wyo.

Also own *a* on right hip.

⛉ on hip and shoulder.

Horse brand, ℺ on left thigh and left shoulder.

RAINSFORD & PALMER.

P. O. address, 149 Broadway, N. Y., Cheyenne, Wyo.

Range, Freezeout Mountain and Little Medicine Bow River, Wyo. Geo. H. Laherty, Foreman.

Also own ⊤ on left hip; dewlap cut down; no earmarks.

Horses branded same as cut on left shoulder.

5

PAGE FROM THE BRAND BOOK OF THE WYOMING
STOCK GROWERS' ASSOCIATION, 1884
Courtesy of the Wyoming State Historical Library

the personnel of those who were to be permitted to engage in the business. Only in a region where the range cattle industry was paramount and where all other economic activities had been subordinated to this one industry could such a system have been maintained. Montana stockmen were never able to get such legal support; indeed, they found it difficult enough to get a suitable inspection law passed to protect them from thieves. The supremacy that the roundup law gave to the Wyoming cattlemen went unchallenged until the grangers arrived in sufficient numbers to insist upon a change.

In order to obtain a legal brand, it was necessary in both the territories to record the same with the county clerk, who was instructed to keep a complete list open to public inspection. This did not meet the conditions of the range-cattle business, for county offices were distant and county lines meant nothing on the range. When an association through consolidation came to be something more than a local organization, the necessity for a general brand book containing the brands of all members was felt. Each year the secretaries prepared such books, containing all the marks and brands for every member and the range occupied. Yearly revision was necessary, for, with the purchase of whole herds, companies added one brand after another to their holdings.[51]

In addition to the necessity of preserving the individual's ownership in the herd and its increase, which resulted in the careful regulations as to brands, roundups, and mavericks detailed above, there was the problem of preserving the individual's control over the blood of his herd. Better bred stock meant heavier steers for market, steers that matured

[51] Copies of the Wyoming Stock Growers' Association Brand Book for each year, 1881-1886, are to be found in the office of the Wyoming State Historian.

earlier and were consequently more profitable. From the early days of the range, certain progressive operators began to import blooded stock from the stock farms of the Middle West.[52] This process of improving the breed fitted in very well with the system of small ranches and close herding. It was genetic rather than extensive. On the open ranges, where there were large profits to be realized from feeding cheap steers on free grass, this attempt to improve the breed was several decades ahead of its time. Wherever the Texas longhorn predominated — an animal that might be packed in its own horns, as the phrase ran on the western plains — those who put money into their herds to improve the blood found their investment threatened. Owners of low-grade Texans enjoyed the advantage of running their cattle on the same range as that occupied by a herd of pure bloods, and the owner of the latter found his increase graded down by the presence of the scrub bulls of his neighbors. Since the latter had as good a right to any part of the public domain as did the owner of a high-grade bull, there was no way to

[52] Good Durham stock was imported early into Montana, where top prices were paid. A prize Durham bull brought $1,885 at an auction in Helena in 1871. (*National Live Stock Journal*, II, 530, December 1871.) The Montana legislature exempted all thorough-bred cattle of American or English registry from taxation for one year after their arrival. (*Laws of Montana Territory*, 1871, Sess. 7, Chap. XC.) Several stock farms were established in Kansas in the early seventies to supply the western demand. The Crane herd in Marion County and the Grant herd in Ellis County furnished good Durham blood for the improvement of the range herds. (McDonald, 81-82.) In the eighties, the Hereford began to supplant the Durham for breeding purposes, particularly on the northern ranges. The Wyoming Hereford Association was organized in 1883 with A. H. Swan as president and George Morgan as general manager. For a time, it maintained the largest herd of registered Herefords in the United States. This concern, which still operates as the Wyoming Hereford Ranch Company, did much to improve the blood of the Wyoming and Montana herds. (John M. Hazelton, *History and Handbook of Hereford Cattle* (Kansas City, 1925), 46.) The importation of Oregon cattle all through the range period and the arrival of eastern stock during the boom were both factors in the improvement of the northern herds.

prevent such an outcome, and the range industry never fully met this difficulty. Democracy, as many critics have noted, turned out to be a leveling-down process.

The older stock growers, particularly in Montana, objected to the fact that the northern ranges were regarded as mere feeding grounds for longhorns from the southern breeding grounds. They held that Montana was not merely the terminus for the "long drive." For years, they had been breeding stock for market of a quality far superior to those from the south. They pointed out that a steer bred on the northern range was a better animal than one from Texas and that he brought a better figure on the market.[53] One of the arguments used against the establishment of a National Cattle Trail from Texas to the northern areas, a

[53] A comparison between the cost of raising a 3½ year old steer in Montana and the same animal in Texas was made in 1881. Charles Goodnight, well-known Texas cattleman, made the estimate for that region and George Clendenning of Fort Benton made the estimate for Montana. The two estimates were published in the *River Press*, December 28, 1881, as follows:

Texas		*Montana*	
Interest on cow valued at $12.50 for one year at 10%..	$ 1.25	valued at $20 at 15%........	$ 3.00
Cost of sirage (graded bull)....	1.50	....	2.50
Holding cow and calf on the range and branding calf....	1.50	....	1.50
Holding steer 2½ years.............	2.50	....	2.50
Taxes for 3 years at 5c per year	.15	at 21c per year....................	.63
Driving to Kansas Market......	.50	to Montana Market.............	.50
Proportion of animal loss estimated on herd	1.35	....	2.00
Total Expense..................	$ 8.75	....$12.63	
Value of 3½ year old in Kansas market	$23.00	in Montana Market...........$28.00	
Net Profit	$14.25	....$15.37	

These figures were for graded Texans and would not apply to many of the herds that in the boom period came up the trail to Wyoming and Montana.

proposition vigorously advocated in 1885-1886 by the southern cattle growers, was that the latter were not mere feeding grounds, but a region where native stock of superior breed was being threatened by low grades from the south.[54] In 1873, a law passed in Wyoming imposed a fine on any owner who allowed a Texan, Cherokee, or Mexican bull, quarter or over, to run at large.[55] During the range period, no such legal remedy was furnished the Montana stock growers, and the problem of clearing the ranges of low-grade bulls was left to the stockmen in their local associations.

In addition to the necessity for some sort of regulation as to the quality of the bulls on the range, the number of bulls per head of cows needed supervision. The herd owner who kept the number of bulls up to a figure sufficient for good breeding purposes found that he was supplying service to those herds on the range that were lacking in breeding stock. At the association meetings, there was much discussion of this question, and resolutions were passed by both the Wyoming and Montana organizations, binding each member to supply one bull for a stated number of cows.[56] Local associations went still further and regulated the time when the bulls belonging to members should be released on the ranges for service. In this way, the calf crop would not come too early in the severe weather and consequently result in a loss.[57] But this question, one of the most important, since it had to do with the number, quality, and vitality of the "calf crop," was never fully solved by the range cattlemen.

The activities of the stock associations were by no means

---

[54] *Rocky Mountain Husbandman*, January 7, 1886.

[55] *Laws of Wyoming Territory*, 1873, Sess. 3, p. 223.

[56] Minute Book, Montana Stock Growers' Association, 95-97; *By-Laws of Wyoming Stock Growers' Association*, 1884, p. 11.

[57] Regulations concerning bulls of the Shonkin Association. *River Press*, December 28, 1881.

confined to the adjustment of matters concerning ownership, the roundup, the maverick, and the breed, questions that arose among the stock growers themselves. The business of raising cattle on the High Plains was exposed to very serious dangers from without. The property of every cattleman was at some time or other threatened by those common enemies, the Indian, the cattle thief, and disease. Here, common action was imperative, and much of the strength of the association was due to the fact that only through very close union and cooperation, could the individual obtain adequate protection against these dangers.

Although the aborigine of the northern plains had by the eighties ceased to be a hostile, he had not yet become a settled, reservation Indian. He was passing through a period of transition, as the old foundations of his existence disappeared and he was forced to seek new ones. Although it is clear to us that he was ceasing to be a major problem in western affairs, the whites who were living in the neighborhood of the reservations were not fully aware of this process of transformation. To them this period of change was peculiarly trying. Pitiable victim of official dishonesty and stupidity as the Indian appeared to the West, he was also an unmitigated nuisance over whom the Government extended an active protection that it refused to the whites. "Against warlike tribes we have, and will again, take our chances," wrote the secretary of the Wyoming Association to the Indian Bureau, "against reservation Indians, acting under official sanction, we claim protection." [58] But the protection that the cattlemen demanded was not forthcoming; indeed the task of carrying the Indian through this period of transition and at the same time protecting the property of the

[58] Sturgis to Price, October 29, 1883, Correspondence of Wyoming Stock Growers' Association, Letter Book, July-December, 1883.

whites would have taxed the powers of a government far more efficient than ours. Since the protection was lacking, the cattlemen acted for themselves.

The problem centered around the question of subsistence. As long as there were any buffalo in eastern Montana and northern Wyoming, the Blackfeet and allied tribes from the reservations north of the Missouri, the Crows below the Yellowstone, and the Sioux from the Dakota agencies were sure to be out looking for them each summer. The Indian Bureau had instructed the agents to issue "passes" to these hunting bands, which were of no particular importance beyond giving official sanction to what would have been done anyway. Once the hunting parties crossed the reservation boundaries, they were practically on their own. It could hardly be expected that these plains Indians, whose independent economic existence had been based almost completely on the buffalo herd, would be very discriminating when they rode by the decaying carcasses of tens of thousands of these animals, slaughtered by white hunters, and saw in their places the herds of the cattlemen moving across the face of the land.

All through the seventies, the Wyoming stock grower had suffered heavily from marauding Indians. Following the Sioux War, 1875-1877, and the clearing of the northern ranges, however, Indian depredations upon Wyoming herds were far less serious than those upon the cattle of Montana. The Shoshone occupied the only reservation within Wyoming and they gave little trouble. In 1881-1883, the stockmen who had rushed their cattle into the Powder and Tongue River country, as soon as it was open, were bothered by hunting parties out from the Sioux agencies in Dakota, and complaints were made to the Indian Bureau. Up on the Tongue was a small band of Northern Cheyenne, who had started to take up land in severalty under the act of 1875. These allotment Indians and the incoming stockmen were soon in difficulties. The former found a spokesman in a squaw man who was assisting them in getting their cabins built and their ranches started. The cattlemen accused the Indians of cattle stealing and cattle killing, and the leader of the Indians countered by writing to the Indian Commissioner that a neighboring cattle company was instructing its cowboys to do everything possible to force the Indians to move out. The accusation was also added that small, white ranchers in the same region were receiving no better treatment than were the Indians.[59]

The reaction was immediate, for, as we shall see, the Association was very sensitive to the charge that it was the foe of the small operator. Post, the Wyoming delegate to Congress, was instructed by the Association to interview Indian Commissioner Price. Post reported that he had done his best to disabuse the Commissioner's mind of the truth of this story but felt that he had failed for "this is the sort of evidence that delights the willing ear and appeals to the sentimental heart of our Christian Commissioner of Indian Affairs." That the large companies were not above using the lawless methods of the frontier for their own advantage, we shall have plenty of occasion to note. The suggestion of the Wyoming delegate to the secretary of the Association was significant.

"I think that if this man . . . . could be taken in on the general roundup next spring and branded with the S & H brand, it might enable his pet Indians to keep track of him when they go on a little cattle killing expedition; perhaps a few printed copies of his letter circulated among the cow-

[59] Copy of the white leader's letter in the Correspondence of the Wyoming Stock Growers' Association, together with letter, Post to Sturgis, February 7, 1883, File, January-June, 1883.

boys might induce them to help civilize his renegade Indians." [60]

The Indians on the reservation in northern Montana gave the stockman the greatest trouble and roused his irritation against the Indian Bureau to the highest pitch. Here was an area of over twenty million acres, embracing some of the best pasturage in that Territory and occupied by possibly twenty-five hundred Indians. By 1881, buffalo and other wild game were so scarce on the reservation that either the Indians must be fed by the Government, or they must go southward into the area between the Missouri and the Yellowstone and east of the Musselshell, where a remnant of the northern buffalo herd was still to be found. [61]

In the spring and summer of 1882, the cattlemen of the Judith and Musselshell regions felt the full force of Indian Bureau inefficiency. Down across the ranges trooped bands of Indians provided with passes from their agent and accompanied by a corporal's guard from a nearby post, a military escort that never saw the majority of its charges after they left the agency. To these, were added other bands from across the Canadian border, who drifted southward in spite of the efforts of the Mounted Police. In the wake of "these breech-clouted pets of the Government," lay the carcasses of range cattle, many of them used as targets to test the skill of the hunters before they reached the buffalo range. [62] Along the Missouri, in the wood yards that had been estab-

[60] *Ibid.*

[61] Memorial to Congress, *Laws of Montana Territory*, 1881, Sess. 12, pp. 131-132.

[62] Letter of Granville Stuart in reply to an editorial in the *River Press,* August 8, 1881. The editor had stated that stockmen were "a little apprehensive" of the damage of Indians during the summer. "They are more than apprehensive," replied Stuart, "they are as scared as hell." The troops were of no assistance. One soldier "wearing eye glasses" was sent out from Fort Maginnis to round up fifty Gros Ventre. *Rocky Mountain Husbandman,* September 4, 1884.

lished to supply the river steamers with fuel, there was the attraction of bad whisky for the simple redskin out on a holiday excursion. Offscourings from the old mining camps, horse thieves, fugitives from the law, and half-breed purveyors of whiskey, all collected in this last rendezvous of the badman of the Montana frontier. These gentry welcomed the young brave, served as "fences" for his stolen horses, and after filling him full of whiskey, turned him loose. Along with his fellows, he showed up on the reservation in the early fall with little or no buffalo meat. Stockmen complained that they fed the Indians all summer, a job the Government should attend to. Back on the reservation, the agent complained that the Indian was lacking in a winter supply of buffalo meat and that the amounts appropriated for supplies by Congress were hardly sufficient to keep him from starvation during the winter.

The contention of the cattleman that the Indian Bureau was hopelessly ignorant of the situation, and that any protection against Indian depredation must come as a result of his own efforts, was fully confirmed by the publication in the territorial press of the correspondence between the delegate of the Territory and the Secretary of the Interior. In reply to an appeal by the Montana delegate for protection, the Secretary wrote as follows:

Upon the subject of the rights of the Indians to leave their reservations for hunting purposes, your attention is respectfully invited to the provisions of the treaty of October 17, 1855 (11 Stats. 637) with the Blackfeet and other Indians. Articles 7, 11, and 12 of said treaty set forth the rights and privileges of the Indians and the settlers, and the remedy for wrongs committed by either. . . . . The occupancy of these lands [Judith and Musselshell regions] is, under the treaty, as much a right of the Indians as of the people of Montana, who are grazing stock thereon and so long as the treaty remains in force, its guarantees should be respected. If this course is pursued by those who are thus keeping their stock on lands not belonging to themselves but to the United States, their safety will be assured and an Indian war . . . . will be prevented. I need not remind you of the fact that while it is the duty of the government to protect the lives of the white man, it is equally its duty to protect the Indian committed to its charge.[63]

Two years later, this Secretary, who was so anxious to protect the Indian, was receiving reports that the Blackfeet were eating the bark off the cottonwoods and poplars along the river banks to keep from starving to death.[64]

The seriousness of this danger and the failure of the authorities to give any protection, led the local organizations in central Montana to get together in 1881 for mutual protection. The stockmen of three separate areas where strong local roundups had developed — the Shonkin, the Mussel-shell, and the Sun-Teton regions — were ready for common action. Small companies were organized, each member of the local associations furnishing his quota of well-armed and equipped riders.[65] The treasurer of each association was directed to post rewards for the apprehension of anyone selling liquor to the Indians or half-breeds, or caught setting prairie fires.[66] During the summers of 1882 and 1883, these small bands of horsemen, detailed from the different outfits,

[63] Letter of Secretary Kirkwood in reply to letter of Major Maginnis in the *River Press*, September 28, 1881. See also Stuart, II, 157-164.

[64] *Annual Report of the Indian Commissioner*, 1884, p. 107. "The Indian is not to blame," declared Granville Stuart before the National Stock Growers' Convention at St. Louis in 1884. "It is the atrocious policy of the United States that is to blame. . . . . They put them on reservations and pretend to feed them and clothe them, but they do not do it. . . . . Hundreds of Indians have perished from starvation in Montana in the last twelve months." *Proceedings of the National Cattle Growers' Association* (St. Louis, 1884), 58.

[65] *Rocky Mountain Husbandman*, August 25, 1881; Letter of Granville Stuart to the Secretary of the Choteau and Meagher Protective Association, in the *River Press*, August 31, 1881, offering the support of the stockmen of the Musselshell.

[66] *Rocky Mountain Husbandman*, September 22, 1881.

Crowd of Indians around the Government slaughter house at the Fort Peck Agency, Montana. This photograph was taken in 1884.

rode the northern ranges as a frontier protective force. Indian camps were broken up and the bands hustled back to the reservation, stragglers were rounded up, and now and then there was a running fight with a troop of young braves out on a horse-stealing foray.

To the cattlemen of the eighties, the danger from the Indian was a passing one, for the Indian himself was passing. The inertia of the Indian Bureau in Washington and the inefficiency and corruption of the force in the field probably worked as well as any conscious policy in bringing about a subjection of the Indian; for it amounted to pacification by starvation and neglect. The younger and more adventurous braves, who, in 1880 and 1881, had ridden out to hunt the buffalo and steal horses as their fathers had done before them, were, by 1887, lining up for their beef issue at the agencies or hanging around the government slaughterhouses. Off the reservation, the Indian was little better than a beggar, who continued to be a nuisance, but never again a serious hindrance, to the economic expansion of the whites.

Serious as the Indian menace was in some quarters, it never threatened to injure permanently the prosperity of the stock growers, as did the threat of the cattle thief. While the problem of the Indian became less and less with every passing year, that of the "rustler" increased as the capital invested in the range-cattle business multiplied, as the means of access to markets improved with the arrival of new railroads, and as the westward-moving farming frontier reached out into the range country, and as communities antagonistic to the prevailing cattle industry round about grew up. Only by constant vigilance on the part of a highly developed organization, operating over all the range, would it be possible to protect the millions of dollars worth of exposed property in a land where the arm of the law was none too strong. The need for mutual aid against the thieves brought the cattlemen into organizations and kept them there, in spite of any conflicting interests. Few frontier groups ever attained the corporate strength and efficiency that some of these cattlemen's organizations displayed, for few needed such constant protection extending over such large areas for such a long time.

Although this last frontier was plentifully supplied with a criminal element made up of drifters from decayed mining camps, hide hunters, and whiskey pedlars, it was not from these groups that the cattle thief came. The experience necessary for his trade was learned only through handling cattle. The worst offenders, and those most difficult to apprehend, were those who at some time or other had been connected with the range-cattle industry.[67] It had been the custom for cowboys to run small bands of their own stock on the range with the herds of their employers. The ease with which a hired hand might add to his herd by branding a few mavericks overlooked in the spring roundup, or by changing the brands of some of his employer's cattle or those of the neighbors, was the cause of many a cowboy's downfall. That the stock growers recognized this danger is evidenced by the passage of resolutions by both the Wyoming and Montana associations binding all of their members not to employ hands who were known to possess any cattle of their own.[68]

[67] The files of the correspondence of the Secretary of the Wyoming Association contain abundant evidence of this. Letters to managers and employers, inquiring as to the character or whereabouts of certain persons, once in their employ and suspected of being cattle thieves, are common. The Executive Minutes contain entries ordering the investigation of members who were guilty of hiring cowboys on the black-lists. See also Will James, *Cowboys, North and South*, 75-79.

[68] The Montana resolution was quoted in the *Rocky Mountain Husbandman*, May 30, 1885; Minute Book of the Wyoming Stock Growers' Association, 102. There was much objection to this, on the ground that it shut out the small owner in order to benefit the large outfits. Territorial newspapers cited this as evidence of the greed of the "bovine-king." (*Helena Weekly*

Once started, the descent was easy. The methods employed by the cattle thief, which have formed the basis for countless romances of the Old West, were curiously unromantic. If the thief had no brand, the first step was to establish one. A brand would be selected and recorded with the clerk in some county, and a small band purchased and branded with that brand. Having established himself as a cattle owner on a small scale, the cattle thief began to build up a herd. Mavericks were picked up and branded when the opportunity offered. Small bunches of likely cattle would be cut out of some herd, usually ranging some distance from the rustler's headquarters, and run off to some secluded spot where the brands could be changed at leisure. The task of altering such brands was much less difficult if that of the rustler was not unlike that of the stolen cattle. The thieves, however, became very expert at the business and were able to produce a changed brand that would defy all but the closest and most experienced examination. Once the brand was on and sufficiently healed, the thief ran the herd in with his own.

The protective measures employed must be adapted to the habits of the rustler. A detective force was scattered out over the ranges to keep an eye on everyone engaged in the range cattle business. No one was above suspicion, least of all the newcomer. It was the stranger with a small herd who received the attention of the detective service, for it was along that road that the cattle thief usually started. Any suspicious action resulted in immediate investigation. The correspondence of the secretary of the Wyoming Association is full of inquiries sent to individuals or to other associations for in-

Independent, August 28, 1885, and *Rocky Mountain Husbandman,* May 30, 1885.) Even among the large owners, many thought it unwise, as it did not encourage thrift among the men but gambling and drinking, since they could not invest their wages in cattle. Minute Book of the Montana Stock Growers' Association, 87.

formation concerning the past history of this man or that. There was a continual shifting of the labor force of the cattle outfits, for the cowboy was a drifter. The associations tried to keep track of this shifting group as well as they could and to warn each other of those cowboys with a record. Each association kept a black-list, which was being constantly revised by a subcommittee of the executive committee. Copies of this list were sent to all members, who were forbidden to employ anyone whose name was so listed. Any infraction of this rule resulted in a demand for a full and immediate explanation to the executive committee. If such an explanation was refused or was insufficient, the disobedient member was expelled. Blacklisted cattlemen and cowboys were also excluded from the association roundups and maverick sales.[69]

In addition to maintaining a careful scrutiny over the personnel of the range cattle industry, the roundups must be closely watched, either by detectives or by association members in whom the secretary had complete confidence and who were in constant touch with the office. This supervision of the

[69] The Wyoming Association acted with promptness and vigor. The following letter from the Secretary to one of the largest cattlemen in the Association is illustrative:

"Mr. James B. Jackson,   July 18, 1885
Council Bluffs,
Iowa
Dear Sir:

"A serious report has been made by the Executive Committee with regard to an occurrence on your roundup during the recent general roundup. It is stated by a member of the Executive Committee and by other responsible parties that your foreman allowed Jack Cooper to accompany your outfit in the Bitter Creek Roundup, that you carried his bedding, furnished him meals, and further that you allowed him to buy mavericks on your own range. You are aware that the Association has regarded Cooper for the past year as a very dangerous man and that the Association is now carrying on a suit with said Cooper for cattle which he stole in 1884. Under the circumstances, to have one of our most prominent members in that section give him aid and comfort in the way above described is a direct injury to the Association

roundups was particularly necessary where, as in Wyoming, the mavericks were the property of the Association. Foremen of Wyoming roundups were given detailed instructions as to the conduct of the roundup and the sale of mavericks, and these were to be followed to the letter. Behind the foreman stood the Association ready to exert all its power to support its representative in the performance of his duty. Any complaint that the foreman had failed to perform his full duty resulted in immediate investigation, and if the charges were substantiated, he was not only dismissed but the bonds, which he was required by law to put up, amounting to $3,000, were forfeit to the Association.[70]

The other side of cattle protection was inspection. Cattle were of no value unless it was possible to get them to market. Sometime, sooner or later, the rustler's cattle got to the railroad or to the Indian reservation or to some community big enough to have a local demand for beef cattle. Only by the inspection of the brands of the cattle offered for sale could the thief be prevented from realizing on his stolen property. This service required a large number of highly trained and honest men scattered over a half-continent. Not only was it necessary to maintain inspectors at the chief loading points in the Territory, but there must be a corps of men at each of the large markets handling the range-cattle business. The

and offsets any effort we can make to protect your property and that of others from this class of men."

The letter closes with a peremptory demand that Jackson appear before the executive committee and make an explanation. Correspondence of the Wyoming Stock Growers' Association, Letter Book, 1885.

In the Minute Book of the executive committee (p. 7) the following entry was made for September 14:

"*Resolved*, in view of the admission of the facts by Mr. Jackson, regret expressed and promise to prevent recurrence, that the charges be laid on the table."

[70] *By-Laws of the Wyoming Stock Growers' Association*, 1884, "Instructions to Roundup Foremen," 16-22.

Wyoming Association maintained exterior inspection at Miles City, St. Paul, Chicago, Clinton, Council Bluffs, and Pacific Junction, and interior inspection at the Pine Ridge and Rosebud Indian Agencies and along the line of the Union Pacific at all the loading stations between Laramie and North Platte.[71] By arrangement with associations in other states and territories, some of the expenses of exterior inspection were shared.[72]

When the cattle were unloaded at the stockyards of Chicago or Omaha, one of the association inspectors was on hand to check the brands of the shipment. It was quite the usual thing to find in a large consignment several brands other than those of the shipper. In making up the full beef herd on the range, a few head that did not belong to the owner of the herd were sure to get in. At the central market, a very careful record was made of these stray brands, and when the money was paid for the cattle by the commission merchant to the owner of the shipment, the price of these strays, minus the freight charges, was deducted, and turned over to the inspector. He in turn sent this money with his record to the home office. The secretary then notified the owners of the stray brands and enclosed the money that the Association had received.

Even if there had been no stealing on the ranges, such an inspection service for association members would have been necessary. But the rustler was abroad in the land, and the members of the Wyoming Association demanded that he be checked through inspecting. Inspectors were instructed to

[71] Minute Book of Wyoming Stock Growers' Association, p. 88, April 2, 1883.

[72] Agreement between the Montana and Wyoming associations whereby the former would do all the inspection at St. Paul and the latter at Chicago and Omaha. The cost of such inspection was apportioned to the two associations. Clay, 263.

hold out cattle with altered brands and to notify the home office immediately so that an investigation as to the origin of the shipment and the identification of the shipper might be started.

These precautions, however, were not enough, for many of the rustlers' cattle had been "mavericked" and arrived at the market with good, clean brands. In attempting to stop this outlet for stolen cattle, the Wyoming Association took a step that brought it into serious difficulties. By a resolution passed November 8, 1883, just prior to the enactment of the Maverick Law making the maverick the lawful property of the Association, it was declared "that all rustlers' brands and all stray brands for which there were no known owners be treated as maverick cattle." [73] Since the Association determined which brands should be regarded as "rustler's brands" and forwarded lists of such brands to its inspectors, it was practically forcing owners who were suspected of being rustlers to prove their innocence to the Association before they could recover the property that had been seized. Such an arbitrary exercise of power was in complete violation of the fundamental constitutional rights guaranteed to the individual. It is not to be wondered at that the Association soon found itself involved in serious legal difficulties and the object of a growing antagonism on the part of town's folk, small cattle owners, and grangers.

Evidence having been found of the existence of fraud and theft, the burden of apprehending the accused and getting him before the bar of justice rested on the Association. Law-enforcing officers of the counties were slow to act, if indeed they were not inimical to the stock growers. Escape was easy in a sparsely settled country, and in the small towns there

[73] Minute Book, Wyoming Stock Growers' Association, p. 102, November 9, 1883.

was always a group ready to extend all possible aid and comfort to the fugitive and keep him informed as to the movements of the officers.

The cattle thief once apprehended, his conviction before a court of law was never obtained without great difficulty. Trials were held in county seats where the crowd, as often as not, sympathized with the accused. Juries were not inclined to be too zealous in the protection of the property of the "cattle barons." Lawyers of great ability found as much profit then in defending cattle thieves as they do today in defending bootleggers. Local officers, whose political fortunes depended upon the good will of the community, were none too active. Unless backed by able and energetic lawyers employed by the Association, they were inclined to let the cases drift. The difficulties of the Wyoming Association are well described by a letter of its secretary written in 1887 to one of the association detectives:

It seems to be the popular thing at present to hamper the work of the Association in every direction. You probably have seen the Cheyenne papers and have learned thereby the result of the many cases tried before our last session of court here. After the expenditure of a great deal of money and our obtaining what we considered ample evidence to convict, the attorney for the defense, the judge on the bench and the jury, as well as the newspapers and the general public seemed combined to place the Association in the light of criminals, with the results that you are probably familiar with.[74]

This task that the Wyoming Association had undertaken of affording complete protection to the property of its members was an enormous one. It involved the expenditure of large sums of money raised by assessments of the members, the sale of mavericks, and dues. It required a large force working out from a highly organized and efficient central

[74] Adams to Smith, January 21, 1887, Correspondence of the Wyoming Stock Growers' Association, Letter Book, 1887, p. 616.

organization. Into the office of the secretary came a constant stream of reports from detectives scattered all over the West, from inspectors at the shipping and unloading points, and from persons in various locations who acted as eyes and ears for the Association. Complaints from members, letters over the question of putting this name or that on the black-list, explanations by members who had been questioned as to some action or other by the Association, communications from railroad officials over freight rates and service, and correspondence with government officials, local and fed-eral — all this volume of association business, touching al-most every phase of life on the High Plains during the eighties, passed through the secretary's hands.

The executive committee was in frequent session. This committee decided the policy of the Association, subject, of course, to the will of the whole body as expressed at its gen-eral sessions. A few excerpts from the Minute Book of the executive committee will give some idea of the nature of the business that came before this body and the power that it wielded:

August 25th, 1885 [p. 4]

Resolved: That an order be printed and sent to all members ordering them to furnish horses, wagons and food when necessary to inspectors in pursuit of criminals or in any emergency. Failure to do so is declared ground for explanation on proof of fact.

November 30th, 1885 [p. 16]

Complaint was made of the failure on the part of John Lind, Fore-man of Roundup No. 6, to properly perform his duty in connection with the Maverick Sales in the General Spring Roundup. The Com-mittee ordered the facts to be laid before Judge Carey, as Bondsman for Lind, with a request that he report on the same to this Committee.

January 23rd, 1886 [p. 20]

It was moved and seconded that the proposed amendment to the Maverick Law as now amended by the Executive Committee be presented to the Legislature.

January 25th, 1886 [p. 21]

The Secretary called the attention of the Committee to the attitude of the *Sun* newspaper toward the Association, as shown by the leading editorial of Sunday's edition. (January 24, 1886.) After full discussion it was moved and seconded that the editor, Mr. E. A. Slack, be requested to appear before the Executive Committee at once.[75]

June 6th, 1886 [p. 42]

On motion [the Committee proceeded] to take up the consideration of the Allen case.

Mr. Corlett, Attorney for the Association, Mr. Allen, Inspector Reese and Detective Parker were then admitted. Mr. Allen made a statement explaining the circumstances under which he tied a calf belonging to Mr. Swan to the underbrush near his ranch and some distance from its mother, and giving the reasons for so doing.

After due consideration of the matter, on motion of A. H. Swan, duly seconded, the Committee decided to exonerate Mr. Allen from the charge of having intended to steal the calf, and dismiss the case with admonition to Mr. Allen that it must not occur again.[76]

July 7th, 1886 [p. 44]

The Committee considered the question of the Governor issuing a new proclamation against the importation of cattle from localities infested with pleuro-pneumonia and of the importation by rail of cattle from localities where Texas fever is likely to originate.

The draft of the new proclamation to be presented to the Governor was prepared, to be again submitted to the Committee, after approval by the attorneys of the Association.

July 20th 1886 [p. 45]

The question of placing the names of the ringleaders in the

---

[75] The *Sun* editor had objected to the proposed law that the license fees collected on saloons and gambling houses in Cheyenne should be turned into the Laramie County treasury. This was interpreted by the editor as a move on the part of the stock growers to relieve themselves of just that much taxes, as practically all the county taxes came from the stock-growing industry. The author of the bill was Teschemacher, himself a member of the executive committee of the Association.

[76] It is interesting to note that Swan at this time was the owner of more than a hundred thousand head of cattle.

"meeting" on roundup No. 23, on the "Black List" was considered and on motion, duly seconded, the Secretary was authorized to prepare and issue the necessary circular, stating the reason for placing said names on the "Black List" to all stock men of the Territory.

The failure of the police power in frontier communities to protect property and preserve order has resulted over and over again in groups who represented the will of the law-abiding part of the community dealing out summary justice to offenders. The history of the American frontier is full of such instances. As long as these self-appointed guardians of life and property are of the frontier and reflect its spirit and methods, they receive general support. As we shall see, the cowboys in eastern Montana ran down horse thieves in 1882-1884 and lynched them, without any serious objection on the part of the community. In Wyoming, the very strength and efficiency of the Stock Growers' Association became its weakness.

Although the latter attempted to punish cattle stealing through the ordinary legal channels, there was good ground for believing that the executive committee was not adverse to more direct methods, and indeed the minutes of the committee show that such suspicion was not unfounded.[77] The idea that a half a dozen wealthy gentlemen, some of them but lately arrived in the Territory, should direct in comparative safety a campaign against those who preyed upon property, the great bulk of which was owned by absentee capitalists, was not pleasing to the frontier mind. The employment of frontier methods by such gentlemen, operating

[77] The Executive Minutes for January 21, 1887, contains the following entry:

"The Secretary read a letter from Inspector Jas. L. Smith, containing an account of the manner in which the expenses claimed by him (in the matter of the killing of John Smith) were incurred. On motion duly seconded, the Secretary was authorized to settle the claim above ref'd to ($100.00)." Executive Minutes, Wyoming Stock Growers' Association, 1887, p. 55.

through a corps of hired detectives responsible only to their employers, resulted in the rapid growth of violent antagonism for the Association on the part of all who were not included in its membership. Town dwellers, small cattlemen, and grangers began to look on the Association as an enemy, its members in control of the chief offices of the Territory, its army of detectives and inspectors at every point, its powerful and active committee all working solely for the advantage of the large cattle growers. Instead of the Territory's being a "cattlemen's commonwealth," as it has been called, an increasing number of people in the Territory were convinced that they were being forced to live under a "cattleman's oligarchy." When the range-cattle industry began to decline, the efforts of the Wyoming Association to retain its power in the face of a rising tide of popular sentiment resulted in the first real struggle in Wyoming politics. The story of this struggle will be left to a succeeding chapter.

In Montana, the herds of the eastern and central sections of the Territory were as much a prey of the rustler as those of Wyoming. Lacking a strong association like that of Wyoming, the Montana cattlemen turned to the territorial government for protection. In 1883, a bill was introduced into the legislature by representatives for the stock-growing areas. This bill provided for a Board of Live Stock Commissioners, five in number, appointed by the Governor. This Board was to employ stock inspectors, whose duty it was to prevent illegal branding, slaughtering, driving, or shipping of stolen stock. These inspectors, who were responsible solely to the Stock Commission, were empowered to make arrests without warrant and to summon bystanders to their assistance, who must help or be guilty of a misdemeanor. To keep up this force, all the taxable property in the Territory was to be assessed a third of a mill each year. The fund thus created was to be known as the Territorial Inspection Fund.[78]

Considerable opposition developed in the legislature to this bill. It was attacked as benefiting only the "cattle kings" at the expense of every one else in the Territory. If they wanted protection over and above what the law gave to every man's property, let them do as the stock growers of Colorado and Wyoming had done, organize and hire their own inspectors. In the legislature, a representative from Silver Bow County, the mining center of the Territory, offered an amendment to the title as follows: "An Act to create the Office of Head Gamekeeper, Chief Firewarden, and Boss Cowboy in the interest and for the sole use of the impoverished cattle owners of said Territory." [79] The minority report on the bill argued "that any interest in this Territory, which pays so high a percentage of profits upon the amount invested as does the stock interest, requires no protection at the hands of the people, but is well able to bear the burdens which are so obviously private in their nature, from its own profits." [80] Although the bill passed the territorial legislature, a newly arrived governor vetoed it on the ground that such a law was a grave menace to personal liberty, a delegation of arbitrary power to unsworn subordinates to make arrests. No legislative benefit likely "to accrue to any stock owner, too negligent to protect his interests under the present civil remedies," would compensate for the additional tax burden.[81] Territorial governors had to be educated.

"Having failed in every endeavor to secure adequate pro-

[78] Copy of the bill found in the *Helena Weekly Herald*, March 1, 1883.
[79] *House Journal*, 1883, Sess. 13, p. 244.
[80] *Ibid.*, 243.
[81] Copy of Governor Crosby's veto message in the *Helena Weekly Herald*, March 15, 1883.

tection," commented the *Rocky Mountain Husbandman,* "the stock interest, which has stood guard over the settlements of the Territory for a decade and a half, must rely on itself to secure what the law guarantees to the rest of the community." [82]

In the summer of 1884, local papers began to carry accounts of sudden forays by bands of heavily armed and well led men down upon the rendezvous of the cattle rustlers along the Missouri. Sharp and deadly clashes occurred. A cowboy's lariat, slung over the branch of a cottonwood, supplied a protection which the Territory dared not grant for fear of favoring a special interest. "Mob law is to be abhorred," wrote one editor, "yet when we consider the great annoyance that the people of the eastern portion of the country have been subjected to, we cannot censure them for thus summarily dealing out justice." [83] "It would be well to understand," wrote James Fergus, who regretted that old age prevented him from active service, "that the hangings and so forth of horse thieves have not been done by bands of lawless cowboys, but was the result of a general understanding among all the cattle ranges of Montana. Sympathizers, who are more or less tarred with the same stick, will be watched and their names put on record." [84] The expenses of the campaign, estimated to have been between $3,000 and $5,000, were paid for by the stockmen of eastern Meagher County.[85]

When the legislature met in 1885, the stockmen were determined to get some legal protection. The necessity of self-protection from Indian and thief had strengthened their local organizations and had made general action on the part of all the stock growers in the Territory possible. At their

[82] *Rocky Mountain Husbandman,* March 15, 1883.
[83] *Ibid.,* August 21, 1884.
[84] *Ibid.,* December 11, 1884.
[85] *Ibid.,* September 11, 1884.

meeting in August, 1884, a permanent organization had been perfected, and they were about to join with the younger association in the eastern part of the Territory.[86] The session of 1885 was known as "the cowboy legislature," a body in which the stockmen greatly outnumbered any other interest.[87]

The law passed by this legislature formed the basis for the legal protection that the Territory, and later the State, has afforded the stock grower up to the present time. In the bill, which was prepared and presented by the stockmen, the executive committee of the Montana Stock Growers' Association was to constitute the Board of Stock Commissioners, with power to appoint inspectors and detectives. In the law as passed, this delegation of the appointive power to the Stock Growers' Association was stricken out, and the governor was empowered to appoint the Board. In the bill, the cost of inspection and detective service was to be met by a tax of two mills on all cattle, horses, and mules in the Territory. Representatives of the agricultural sections, where the small farmer could see no benefit accruing to him from such a law, voted against it. As a result, the bill was amended so that the tax was levied only on those counties named in the act, counties where the range industry was paramount, namely, Dawson, Custer, Yellowstone, Choteau, Lewis and Clark, and Meagher.[88] With the passage of this law, the stockmen in Montana were relieved to a considerable degree from the burden of inspection and detective service. Although cattle rustling did not stop, as the Annual Reports of the Board of Live Stock Commissioners show, an efficient

[86] *Supra*, p. 124.

[87] *Bozeman Chronicle*, January 28, 1885; *Great Falls Tribune*, May 28, 1885.

[88] *Laws of Montana Territory*, 1885, Sess. 14, pp. 91-95. A copy of the bill as prepared by the stockmen is found in the *Rocky Mountain Husbandman*, February 19, 1885.

system had been devised within the legal framework of the territorial government, which kept it down to a minimum.

The efforts of the Montana cattlemen to obtain protection had been limited by other strong and independent interests in the western section, which had prevented them from getting all they had asked. Their demand that the government take over the task of inspection and protection through a Board of Live Stock Commissioners was evidence that their organizations were not strong enough to do it unaided. The limitations put upon them as to the appointment of the board, and the placing of the burden of paying the cost of protection on the range cattlemen of the central and eastern sections rather than on the whole Territory, is a fair measure of the importance of the range-cattle industry in Montana as compared with Wyoming.

Although the small stockman and the range cattleman might differ as to the methods of protecting their stock from thieves, they could come together in demanding that the local government take some action to protect their property from disease. Texas cattle, particularly from the southern section of the state, were more than likely to be infected with the so-called "Texas fever," to which they seem to have been partially immune. The danger of allowing Texas stock to come in contact with other cattle was recognized as soon as the drives began, and measures were taken in the neighboring states and territories to protect the local herds. It was found that Texas cattle that had wintered outside of Texas or had been driven north in the late fall or early spring were free from this parasite. Because of this fact, early legislation took the form of prohibiting the introduction of such cattle during the spring and summer, or of providing quarantine areas for those arriving during these periods.

The first state to take legislative action against the Texas fever was, naturally enough, Kansas. In 1859, the legislature of that Territory passed an act prohibiting the driving of Texas, Arkansas, or Indian cattle into the four eastern counties, where they might come in contact with local stock, between June and November.[89] As the Kansas farmers moved westward, the legislature extended the prohibited areas, cutting down the western section where cattle might enter during the forbidden period, there to be held in quarantine until the late fall before they could continue. Drovers with herds for the northern ranges were forced either to bend the line of the drive further and further westward to escape these restrictions, or plan to start in time to be out of the state before the quarantine became effective.

Similar laws against Texas cattle were passed by Missouri, Colorado, Nebraska, and Dakota. Neither Wyoming nor Montana took any early action against the Texas fever, as it was felt that the long drive from the southern grounds and the quarantine regulations maintained in the intervening states, were sufficient to prevent the introduction of disease.

These early laws assumed that all Texas cattle were infected, and therefore must be subjected to restrictions and regulation. Since there was no discrimination, the state was practically regulating commerce in cattle between itself and Texas. Such an exercise of power was clearly in violation of the Federal Constitution. In a case brought up to the Federal Supreme Court over the constitutionality of the Missouri quarantine law, the court held that the state in order to be wholly within its jurisdiction in the exercise of the police power, must provide for expert inspection, which would see to it that quarantine restrictions were laid only on diseased animals or those suspected of disease. Without this inspection, such a blanket restriction on the commerce of one state

[89] *Laws of Kansas Territory*, 1859, Sess. 5, p. 622.

with the others was a usurpation of the power of Congress to regulate commerce among the several states. The court, therefore, declared the law unconstitutional.[90]

The practical result of this decision was to force the local governments to provide some sort of veterinary service if they were to protect themselves from infected cattle arriving from without. The outbreak of pleuro-pneumonia in the eastern states and the increase in the shipment of cattle by rail were additional incentives. As a result, laws were passed providing for an inspection system by veterinarians employed by the local governments in Kansas 1882-1884, Wyoming 1882, Nebraska 1885, Colorado 1885, Montana 1885.

The northern stock growers were particularly alarmed, for just at this time hundreds of eastern cattle, some of them from regions where pleuro-pneumonia had broken out, were being shipped out to the Montana and Wyoming ranges. In addition to the danger from pleuro-pneumonia, Texas fever, which had not caused any serious alarm in the north during the early days of the industry, now became a real menace as the practice of shipping Texas cattle by rail became more and more common. Secretary Sturgis of the Wyoming Association pointed out in his annual report in 1884 that of the one hundred thousand Texas cattle contracted for by Wyoming and Nebraska cattlemen for that year, a considerable part would travel as far as Ogallala by rail. He insisted that some method of proper quarantine for such rail cattle be established or the whole northern industry would be seriously threatened. He disclaimed any intention on the part of the northern growers to injure the Texas trade, but warned the Texas drovers that if the fever broke out on the northern ranges, it might stop the Texas drives altogether.[91]

[90] Railroad *v.* Husen, 95 *U. S.* 465, October, 1877.
[91] *Cheyenne Daily Sun*, April 8, 1884.

That his fears were not unfounded, was proven by the reports that came into his office during the summer of 1884. Herds of five thousand head or more of longhorns came trailing into Wyoming from Sydney and Ogallala, infected with Texas fever and losing thirty to a hundred head daily as they moved northward across the crowded ranges, threatening with every mile the herds thereon. Local owners were powerless and appealed to the Association to get the Governor to establish a quarantine.[92] The result was the issuance of quarantine proclamations by both the Wyoming and Montana governors against all Texas cattle coming upon the northern ranges by rail, until such time as they could be inspected and declared free from disease by their veterinarians.

As Secretary Sturgis had stated, it was the ease and speed with which livestock could be moved by rail and the increasing amount of such interstate movement that constituted the real danger. Cattle infected with pleuro-pneumonia in New Jersey or Pennsylvania were nearly as serious a threat to the cattle of Wyoming and Colorado, as they were to Ohio or Maryland livestock. Because of this, range cattlemen were willing to combine with the feeders and dairymen of the Middle West in demanding that the Federal Government act so as to confine the plague to the areas infected.

The weakness of the states and territories in trying to control the spread of cattle diseases was patent. In the first place, any real local restriction would result in the state finding itself in that debatable zone existing between the police power of the state and the power of the Federal Government over interstate commerce. Again, it was obviously more effective to establish a quarantine over the

[92] Letters, Frewen to Sturgis, Brewster to Sturgis, and Oelrichs to Sturgis, Correspondence of Wyoming Stock Growers' Association, Letter File, May-August, 1884.

area where the disease had broken out, than it was for all the rest of the country through its separate state and territorial governments to set up barriers against the admission of cattle from the infected area. Stock growers urged that it was a legitimate exercise of Federal power for the Government to declare such area quarantine upon the advice of scientific experts in its employ.

The Federal Government might have hesitated in using its power over interstate commerce for the suppression of disease, if it had not been forced to act by circumstances arising in another quarter. Early in 1879, Secretary of State Evarts was notified by Thornton, the English minister in Washington, that pleuro-pneumonia had been discovered among a shipload of cattle sent from Portland, Maine, to England.[93] Evarts' position in the matter was particularly weak, for we had no proper inspection of export cattle, so that he was forced to assume that this was an isolated instance and that the cattle had possibly contracted the disease on shipboard due to the crowded conditions and bad ventilation of the cattle boat.[94]

There was, as a matter of fact, no inspection worthy of the name. In 1878, the State Department got word from our minister to England that that Government was contemplating a restriction on American cattle. As a result, the Secretary of the Treasury issued, December 18, 1878, a circular authorizing the port collectors to inspect American cattle for export and to issue health certificates. Shippers, however, were not compelled to submit to such inspection.[95] This

---

[93] Great Britain, *Parliamentary Papers*, 1878-1879, Vol. 58, "Correspondence connected with the detection of pleuro-pneumonia among cattle landed in Great Britain from the United States of America," Salisbury to Thornton, English minister in Washington, January 30, 1879, p. 1.

[94] *Ibid.*, 2, Thornton to Salisbury.

[95] *Synopsis of Decisions*, Treasury Department, 1878, Circular No. 3823, p. 750.

amounted to no inspection at all, as the port collectors were
wholly untrained in veterinary matters even if the owners
had asked for inspection. On February 1, 1879, Sherman
issued another circular ordering collectors to inspect all
cattle and instructing them to refuse to pass any stock without a certificate of freedom from disease.[96] This was not
sufficient and ten days later the English Government issued
an Order in Council, requiring that all American cattle arriving in England must be slaughtered at the port of debarka-
tion within ten days after their arrival.[97] On a request by the
American minister that the order be modified provided that
the American Government took steps to furnish adequate
inspection at the port, the English Government replied that
it was the feeling of the Council that no inspection at an
American port, no matter how perfect, offered complete
security to the British stock grower.[98]

As a result of the action of the British Government,
which amounted to an embargo on a very profitable trade,
the Federal Government began to stir itself. On March 3,
1881, Congress appropriated $15,000 to enable the Secre-
tary of the Treasury "to procure information concerning
and to make inspection of neat cattle shipped from any port
in the United States to any foreign ports, so as to enable
him to cause to be issued to the shippers of such cattle certificates showing in proper cases that such cattle are free from
the disease known as pleuro-pneumonia." [99] About all this
meant was that the Secretary of the Treasury was to try to
do something about pleuro-pneumonia. The succeeding Congress appropriated $50,000 in each of its two sessions "to
enable the Secretary of the Treasury to cooperate with

[96] *Ibid.*, 1879, Circular No. 3867, p. 32.
[97] Great Britain, *Parliamentary Papers*, Vol. 58, pp. 6-7; *Report of the
Secretary of the Treasury*, 1880, p. xxxiii.
[98] *Ibid.*, 14.
[99] 21 *U. S. Stats.* 442.

state and municipal authorities, and corporations engaged in cattle transportation by land and water, in establishing regulations to prevent the spread of pleuro-pneumonia and to establish quarantine stations." [100]

A commission, commonly known as the Treasury Cattle Commission, was appointed and in 1882, brought in its report.[101] The report covered the whole problem of the suppression of cattle disease, putting the matter of a proper inspection of exports where it belonged as merely a part of the general question. It urged that the Federal Government take action on grounds many of which have become familiar to the student of the extension of Federal power. Against a common enemy, Federal action was justifiable and the expenditure of funds in such a cause should come from all. Federal inspection would result in uniformity of action impossible under the existing circumstances, where the precautionary measures of one state might be nullified by the laxity of its neighbors. Finally, it pointed out that the British Government was not satisfied with state inspection and insisted that Federal inspection be substituted, if American cattle were to be allowed entrance into English ports. In closing, it recommended

. . . . such action on the part of Congress as will confer on our commission or upon some department of the government, the authority to prescribe rules and regulations under which the sound cattle of any State or Territory or of the District of Columbia found infected with lung plague, may be transported or taken therefrom, and under which healthy cattle may be transported or taken through such States, Territories or District; and to provide bonded yards and quarantine.

As to the enforcement of such regulations, it advised that

[100] 22 *U. S. Stats.* 313, 613.

[101] Report of the Treasury Cattle Commission on the lung plague of cattle or contagious pleuro-pneumonia. *Sen. Ex. Doc.* No. 106, 47 Cong., Sess. 1, 1882.

". . . . when officially promulgated [they] should have all the force of law, and such penalties should be provided as may be necessary. . . . ." The action thus advised would make it possible ". . . . to give a clean bill of health to cattle for export . . . . the object for which we were appointed." [102]

The stock growers of the country had an opportunity to express their desires for Federal protection when Commissioner of Agriculture Loring issued a circular letter to all stock growers to meet in November, 1883, at Chicago while the Fat Stock Show was in progress there.[103] In his call, the Commissioner urged the stock growers to come together to consider the extent of contagious disease in the United States, the modes by which it had been introduced and disseminated, the methods of eradication, and the legislation necessary to bring this about. On November 15-16, 1883, one hundred and seventy-five delegates, representing twenty states and territories met in Chicago. The majority of the delegates were from the feeding and dairy states of the Middle West, the range-cattle industry being represented as follows: Wyoming 16, Colorado 6, Arizona 1, and Texas 1. The preponderance of Wyoming delegates in the representation from the High Plains was due to the strength of the Association of that Territory, the active interest that it had taken in the prevention of disease, and the feeling on the part of its members that only through Federal assistance could the combined dangers of Texas fever and pleuropneumonia be averted.

In order to get immediate action by Congress and to assure that such action resulted in suitable and adequate legislation, the convention appointed a legislative committee of eight, who were to meet in Washington in the following January to consult with the Treasury Cattle Commission and

---

[102] *Ibid.,* p. 82.
[103] *National Live Stock Journal,* XIV, 448, October, 1883.

the Commissioner of Agriculture. They were instructed to "suggest to Congress such points of legislation as they may deem best calculated to protect our interests and remove foreign prejudices against our meat production." Thomas Sturgis of Wyoming, who had been made secretary of the National Convention, was chosen as one of this legislative committee.[104]

Against the action of this first National Convention of Stock Growers, opposition developed in two quarters. The fear that the creation of a Federal quarantine would stop the Texas drives and the shipment of Texas cattle to market brought the cattlemen of the Southwest around as a solid unit against the proposed legislation. To them, the action of the Chicago convention, in so far as it concerned the western range, was the result of the determination on the part of the Wyoming men, so conspicuously represented, to shut out southern cattle from competition with the northern stock. The southerners were perfectly willing to act against pleuro-pneumonia, but as to the Texas fever, if it existed at all, it was only in the mind of the northern stock grower. The attitude of the Texas people toward the whole matter was well expressed by a Texas correspondent, writing in the *Breeder's Gazette*:

Why can't we that are interested in the meat production of the United States all work together to get Congress to destroy the pleuro-pneumonia in this country and prevent the possibility of its being introduced, without trying to protect breeders in Wyoming from the competition of a portion of the country more favored by nature? Let each State decide for itself whether it will pay them best to handle the healthy Texas-born cattle or the more expensively raised animals that are liable to splenetic fever. Whatever laws may be

[104] *National Live Stock Journal*, XIV, 533-534, December, 1883. Clay, 120-121.

passed, the extension of the infected area and the cheap price at which we can raise animals will settle it in favor of Texas.[105]

The attack on the Texas longhorn brought to its defense allies from another quarter. The Chicago stockyards and commission men and the packers were of no mind to see the competition between the northern and southern cattle destroyed by an embargo on the latter. Such an eventuality, they feared, would place them at the mercy of the northern stock growers. Among the petitions that came into Congress in the winter of 1883-1884, urging Congressional action, was one from Chicago signed by cattle brokers and commission men against such legislation. This petition declared that no contagious disease was known to exist west of the Allegheny Mountains and that a country-wide inspection was unnecessary and was opposed by all practical stock raisers, feeders, and dealers. It was pointed out that in order to carry into effect the work contemplated, the Federal Government would turn loose on the country a horde of officers, "horse doctors," interested only in finding what they were sent out after, disease. All this would result in needless expense to the Federal Government.[106]

From 1879 on, bills were introduced in Congress providing for some sort of Federal agency to check the spread of cattle disease. In the first session of the Forty-seventh Congress, a bill to create a Bureau of Animal Industry passed the House but failed in the Senate.[107] In the next session, the bill came up again, and, after considerable change, was passed and approved by the President, May 29, 1884.[108]

[105] *Breeder's Gazette*, V, 989, June 26, 1884. State pride forbade Texans to call this ailment anything but the "so-called splenetic fever."

[106] *Breeder's Gazette*, V, 223-224, February 14, 1884.

[107] *Congressional Record*, 47 Cong., Sess. 1, 1881-1882, pp. 381, 5113-16, 6827-30.

[108] 23 *U. S. Stats*. 31.

The bill as finally passed created within the Department of Agriculture a bureau to be known as the Bureau of Animal Industry, headed by a competent veterinary with a clerical force not to exceed twenty. The Bureau was permitted to make rules and regulations to suppress cattle disease and to invite the states and territories to cooperate in carrying them out. If any state or territory accepted this invitation, the Bureau might expend some of the $150,000 appropriated to aid that section in extirpating the disease. In the original bill, the President of the United States was empowered to declare a state or territory quarantine, whenever the said state or territory refused to cooperate with the Commissioner of Agriculture in extirpating disease within its boundaries. This part of the bill was stricken out by the House, the only power remaining being that of the Commissioner to prohibit the transportation of livestock which were actually diseased.[109] In the Senate, the Texas senators succeeded in legislating out of existence the Texas fever by adding a proviso that "the so-called splenetic or Texas fever shall not be considered a contagious, infectious, or communicable disease within the meaning of the sections 4, 5, 6, and 7 of this act, as to cattle being transported by rail to market for slaughter when the same are unloaded only to be fed and watered in lots on the way thereto." [110]

In the debate on the bill, all the arguments of states rights were invoked. Enlightened self-interest would do what a horde of "horse doctors" could never hope to accomplish through a dangerous extension of the Federal power. Texas senators and representatives declared over and over again that Texas fever did not exist save in the minds of veterinarians and their allies, the northern stock growers. In the vote on the bill in the House, out of the 127 voting against

---

[109] *Congressional Record*, 48 Cong., Sess. 1, 1883-1884, pp. 899, 1465.
[110] *Ibid.*, 3461, 3536.

the bill all but two were Democrats. The bill passed through the support given it by the northern dairy and feeding states.[111]

There was much resentment by the sponsors of the bill against those whose opposition had resulted in a law that fell far short of the vigorous action by the Federal Government necessary to protect the cattle grower. It was felt that the interests of the majority had been sacrificed to those of the Chicago buyers and packers. The *Chicago Tribune*, commenting on the law as it finally got through Congress, remarked,

If the cattle raisers of Wyoming, whose herds represent $100,000,000, really supposed that their vigorous assertion of the necessity of a cattle quarantine on the grounds of public policy would have any effect on Congress, they are undeceived by this time. They and the public interests are too far off to be heeded by Congress.[112]

That the bill passed at all was in part due to the fear that the Chicago interests had of losing the good will of the Wyoming Association. The petition to Congress against the bill which was circulated in the Chicago yards and signed by many of the commission men, was bitterly resented by the Wyoming men. At the meeting of the Wyoming Association on April 7, 1884, all the signers were blacklisted. This vigorous action had its effect. In a letter in the files of the secretary of the Wyoming Association from a member who was in Washington lobbying for the bill, the latter gave an account of a conversation with certain persons from Illinois, who were the chief opponents of the bill. He stated that they had ceased to oppose the bill on the ground that "the Chicago men could perhaps secure some Wyoming shipments which they are at present very liable to lose." [113]

[111] *Congressional Record*, 48 Cong., Sess. 1, 1883-1884, p. 1465.
[112] *Chicago Tribune*, May 27, 1884.
[113] G. W. Simpson to T. Sturgis, May 10, 1884. Correspondence of the Wyoming Stock Growers' Association, File, May-August, 1884.

The scare over pleuro-pneumonia and Texas fever and the discussion as to the proper measures necessary to protect the cattle grower were followed by public uneasiness as to the kind of meat that was issuing from the large packing houses. If diseased cattle were a threat to every stock grower, diseased meat was no less a menace to every consumer. The action of European governments in prohibiting the importation of American meats on the ground that no proper inspection was maintained at the plants where meat was prepared, gave additional weight to the demand that the Federal inspection be extended from that of cattle in interstate commerce to that of meat products as well. As a result, the work of the Bureau of Animal Industry was expanded in 1890 by assigning to it the task of meat inspection. Nor did the Federal Government stop there, for with the increase in the preparation of food by industrial processes outside of the home, there was need for extending Federal inspection so as to include all such food. The passage of the Pure Food and Drug Act of 1906 was a further enlargement of this function of salutary inspection.

This extension of the Federal power in the field of inspection is of great significance, for it has brought into the civil service a new type of government servant. Heretofore, the amount of specialized scientific knowledge required by the Government in order to function was small. Outside of the Army and Navy Departments, the Government operated with a minimum of scientific knowledge and a maximum of clerical routine. Inspection, in order to be worth while, demanded the employment of trained veterinarians and chemists. Since such persons could not be obtained through the old methods of political preference, the Government was forced to develop some machinery whereby such posts would be filled by those trained to the task. The reform in the civil service, contemporaneous with these initial steps in the field of salutary protection, created the means necessary for making a proper selection.

Finally, these new functions of government contained within themselves possibilities for enormous expansion. The government laboratory and experiment station where the scientist finds no limits to his possible activities, save those set by the boundaries of human knowledge itself, have been an ever present force in widening the field of government activity.

VI

THE CATTLEMAN AND THE PUBLIC DOMAIN

The struggle over the passage of the law creating the Bureau of Animal Industry brought out the sharp division of interests growing up between the northern and southern stock growers. The connection that had been established in the seventies between the Texas stock region and the newly opened northern ranges had become the outstanding feature of the range cattle business of the High Plains. Year after year, as the herds came up over the trail, the northern cattlemen were out to buy more stock with the money they had received for their beef shipments. Texas longhorns were cheap, the supply appeared limitless, and they were sufficiently matured so that they could be turned off after a season or two of fattening on the northern grass.

Although this traffic had increased during the northern cattle boom, 1879-1885, and had been of immense profit to the Texas growers, who had sent northward each season greater and greater trail herds, a situation was developing that made the Texas cattleman uneasy even in the midst of his prosperity. In the first place, the shipment of young stock from the Middle West by hundreds of thousands suddenly crowded the northern ranges, which might have continued an outlet for Texas cattle for years to come. Second, the northern cattleman was beginning to pay more attention to breed, for as a breeder of better and heavier stock, he believed he could make more money than as a mere fattener of Texas longhorns. Better breed meant an increased value of the stock and a correspondingly greater risk. To every northern cattleman who was buying better bred stock, the Texas cattle were becoming a menace, which not only threatened his improved herd with disease and inferior blood but with famine as well.

Not only was the character of the northern range cattle business undergoing a change which, in time, was certain to contract the Texas cattleman's market, but there was also the probability that unless some action were taken, the cattle trail along which for more than a decade the herds had gone north would soon be a thing of the past. The quarantine regulations of Kansas were becoming more and more of a barrier, and the westward-moving farmer, with his lines of fence, narrowed down the regions through which Texas cattle could be driven.[1] In addition to this, the cattlemen in western Kansas, finding their range disappearing as the result of overcrowding and the extension of settlement, began to look with disfavor on the trail herds that crowded in among their cattle during the quarantine period and ate off the new grass. Kansas cattlemen at a meeting of the Western Kansas Stock Growers' Association at Dodge City in 1884 declared, "The time has now come when an entire change of policy should be inaugurated regarding the driving and holding of through Texas cattle over and upon the ranges occupied by members of this Association," and urged that the line which had been drawn in a north and south direction across the state, east of which no Texas cattle could be driven, be moved westward to the western boundary of the state.[2]

If the northern outlet were closed, Texas drovers would be put out of business, and stock growers would be wholly dependent on the railroads and the eastern packing centers.

[1] In Kansas, 1884, a special session of the legislature, called for the purpose of stamping out cattle disease in the state, had passed a law, creating a Sanitary Live Stock Commission and had narrowed the quarantine ground for the seventh time. *Laws of Kansas*, Special Session, 1884, pp. 5-13.

[2] *Breeder's Gazette*, V, 584, April 17, 1884.

The thousands of young stock that had been raised for the northern ranges were likely to be a total loss unless their owners could get them north, for they could not be sold in the Chicago market.[3] The excitement in Texas caused by the rumor which got about in the winter of 1883-1884 that the trails through the Indian Territory were going to be closed by the Indian Bureau, and the relief when it was learned that the trails would be open as usual, is evidence of the seriousness of the situation confronting the Texas grower.[4]

The example of the northern range cattlemen and middle-western stock growers in getting together in Chicago in 1883 to urge the passage of the Animal Industry Bill was not without its effect on the Southwest. The northern group had been successful in enlisting the aid of the Federal Government to further their interest; similar action might induce Congress to get the Texas stock grower out of his difficulties.

Early in 1884, the southwestern cattlemen began to agitate the calling of a convention of all the stockmen in the country, a meeting which, it was hoped, would be more representative than the Chicago body had been. A call was circulated in January throughout the country urging all stockmen to join in forming "an organization representative enough and powerful enough to secure advantages and recognition which are now denied us, but which are accorded railroad, manufacturing companies, and other branches of trade. Such interests," the promoters urged, "are guarded and recognized by a national protective system, secured only by national organization, such as we now propose for adoption to the cattlemen of this country."[5] From the signers of

[3] Clay, 119.

[4] Exchange from the *San Antonio Express,* quoted in the *Breeder's Gazette,* V, 313-314, Feb. 28, 1884.

[5] *Proceedings of the National Cattle Growers' Convention,* 1884 (St. Louis, 1884), 1-2.

this call, it is clear that the movement had its origin in the Southwest.

The nucleus for a national organization had been created by the meeting held in Chicago in 1883. It was hoped by those who had been active in creating this first association, that at its next regular meeting, set for the fall of 1884 in Chicago, the southern stock growers would join. But the action of the Chicago convention in supporting the Animal Industry Bill, so bitterly opposed by the Southwest, made it very doubtful that such a union could be brought about.

The date set for the Chicago convention was November 13-14, 1884. The Southwesterners decided to hold their meeting in St. Louis on November 17-18. It appeared to many that the St. Louis crowd were attempting to jockey the Chicago group out of their meeting and force them to join the St. Louis meeting, where they would be in a hopeless minority.[6] The Chicago meeting was held, however, and after much discussion, it was voted to join the St. Louis convention provided that such action "in no case nullified or changed the organization perfected in this city or officers elected." [7]

On such an unstable basis, the two groups met in St. Louis, November 17, 1884. The crowd of twelve hundred assembled there never got beyond the mass-meeting stage, although a constitution was written and officers elected. The cattlemen of Wyoming and Montana were represented by eighty and forty delegates respectively, a small minority as compared with the southwestern representation of three hundred and forty from Texas, sixty-six from New Mexico, eighty-three from the Indian Territory, and fifteen from Arizona.[8] After two days of speech-making and wrangling,

[6] *Breeder's Gazette*, VI, 153, July 31, 1884; VI, 385, Sept. 11, 1884.
[7] *Ibid.*, VI, 749, Nov. 20, 1884.
[8] *Proceedings*, 76-77.

the convention got through a resolution urging Congress to establish a National Cattle Trail from Texas to the northern ranges.[9] It was proposed that a strip of the public domain, six miles wide, extending from the Red River to the Canadian boundary, be set aside as a permanent quarantine ground over which southern cattle could move unhindered to the northern areas. It was urged that the central government, which had turned over millions of acres to the railroads, could hardly refuse "to donate some to the people in the interest of cheaper food."[10]

The Wyoming and Montana delegations bitterly opposed the resolution and were generally supported by those groups that had made up the Chicago association. Because of the scant consideration they received at St. Louis, the Chicago crowd withdrew on the second day of the convention and issued the following statement:

> Finding that harmonious action was impossible and being unwilling in any way to commit themselves to the proceedings of a convention, confessedly sectional in character, the representation of the National Cattle Growers' Association withdrew.[11]

In spite of all the excitement that the St. Louis convention caused throughout the western country and the talk for and against the National Trail, there was little or no chance that the wishes of the southern cattlemen would be realized. Three important obstacles were in the way: first, the Kansas quarantine laws; second, the western farmer, for, as John Clay puts it, "the Kansas granger, the Arkansas traveler, and the Russian settler had their eye [*sic*] on western Kansas"; and third, the crowded condition of the northern ranges.[12] The attitude of the northern stock grower was best expressed by the following squib in a Montana paper:

---

[9] *Ibid.*, 88-89.   [10] *Ibid.*, 15.
[11] *Breeder's Gazette*, VI, 901, Dec. 18, 1884.
[12] Clay, 120.

Nearly every Montana stockman is opposed to the National Trail. 'Cause why? We-uns just got pie enough to go around, and ain't got none to spare for you-uns. See? [13]

The Montana and Wyoming opposition to the National Cattle Trail was due chiefly to the fear of overcrowding. The basis of the range cattle business was the free pasture of those unoccupied and unused leagues of the public domain open to all — the big cattleman and the owner of only a few head, the granger, the sheep herder, the trail herder, the owner of well-bred cattle, and the cattleman with the worst of scrubs. On it, one man's right was as good as another's and all must take their chances. In the opinion of many cattlemen, this basis, none too secure as compared with privately owned pastures, was being seriously threatened by the booming of the cattle business. Writing in 1883, just when enthusiasts were declaring that there was no limit to expansion, a Montana editor observed:

The cattle interests in Montana are threatened with imminent danger from two sources. One is the overcrowding of the ranges and the other, the shipment of diseased cattle from the States. For these many years, the cattle interest has run easy . . . . and the mishaps and dangers . . . . have been of a temporary nature. . . . . But a time has at length arrived, when the herd owner can see in the near future mountains of difficulties to combat. As civilization has progressed, our large cattle herds have kept to the front and of late years, have sought those regions not adapted to farming. But the increase of herds and the continual driving in of stock . . . . begins to tell but too plainly that the overcrowding of our ranges is only a question of a very few years, unless some plan is hit upon to prevent it.[14]

The future safety of the range industry depended upon the success of the stockmen in hitting upon a workable plan to protect it. From the Federal Government, owner of the

[13] *Bozeman Chronicle*, Feb. 18, 1885.
[14] *Rocky Mountain Husbandman*, Sept. 6, 1883.

public domain, little or no help could be expected. The stock grower understood that he was a tenant by sufferance.[15]

It is admitted that every one of the 49,000,000 inhabitants of the United States owns an equal share of, or interest in, every spear of grass on the public domain [declared James Fergus in a speech before the Montana Association in 1879], but few of them have any desire to take immediate possession of the same — too many obstacles stand in the way; they see too many Indians about those particular spears and they say, "Well, after all, I believe I shall not bother with mine just now, you may have it if you want it." . . . . The trouble is, how are you going to divide it?[16]

It was the division of these spears of grass after the Indian had ceased to hedge them about that constituted the real problem of the cattleman's frontier. "The business was a fascinating one," writes Granville Stuart in retrospect, "and profitable so long as the ranges were not overstocked. The cattleman found ways to control the other difficulties, but the ranges were free to all and no man could say with authority when a range was overstocked."[17]

In the early days of the cattle range, the custom of priority, the system of "squatter sovereignty as old as the frontier" was enough to meet the situation.[18] There was room enough for all, and when a cattleman rode up some likely valley or across some well-grassed divide and found cattle thereon, he looked elsewhere for range. The early laws of the states and territories, providing punishment for those who drove stock from their "accustomed range," although primarily directed against stealing, incidentally recognized the fact that by grazing a certain area, the stock

[15] Joseph Nimmo, *The Range and Ranch Cattle Business of the United States* (Washington, 1885), 6.

[16] *Rocky Mountain Husbandman*, March 20, 1879.

[17] Stuart, II, 185.

[18] *Cheyenne Live Stock Journal*, quoted in the *Breeder's Gazette*, VI, 85, July 17, 1884.

grower was in the way of gaining a kind of prescriptive right to the same as over against the newcomer who should attempt to drive off the stock thereon.[19] Because the Diamond J cattle were accustomed to range along a certain creek, that area came to be known as the Diamond J range.

As long as there was plenty of room for expansion, the principle of priority would serve. "Our good luck consists more in the natural advantages of our country than in the scale of our genius," wrote one pioneer Montana stock grower, with unusual discernment as to the importance of the frontier in American prosperity. "Those natural advantages were simply the immensity of our ranges and the facility of locating our herds at a respectable distance from each other. . . . . Those advantages are gradually disappearing, and while it is true that there are a great many locations to be taken up yet, there is no more a vast scope of country at the disposition of stock growers." [20]

The idea that a certain area might become the accustomed range to be held against all comers on the basis of priority right was developed to its greatest extent during the early days of the boom period. Cattlemen, hurrying into the regions of eastern Montana and central Wyoming, set up "claims" to certain areas in the same manner as the miners had done. The newspapers of that section ran long columns of claim advertisements. For example:

I, the undersigned, do hereby notify the public that I claim the valley, branching off the Glendive Creek, four miles east of Allard, and extending to its source on the South side of the Northern Pacific Railroad as a stock range. — Chas. S. Johnson.[21]

Such a claim was, of course, unenforcible in any court of

19 *Laws of Texas*, 1866, Sess. 11, pp. 187-188; *Laws of Wyoming Territory*, 1873, Sess. 3, p. 225; *Laws of Montana Territory*, 1872, Sess. 7, p. 287.

20 Letter of Chas. Anceny to the *Rocky Mountain Husbandman*, Feb. 9, 1882.

21 *Glendive Times* (Glendive, Mont.), April 12, 1884.

law. In order to exclude intruders from the valley to which he had laid claim, Johnson must either get control of all the water adjacent to his range, or he must unite with his neighbors in the same district in refusing to cooperate with any newcomer.

The first of these alternatives depended upon the physical environment. It was an axiom of the "cow country" that water controlled the range. Cattle were able to travel many miles to water, but water they must have, and he who had title to the land upon which there was water was not likely to be troubled by outsiders crowding in on the contiguous range. Under the land laws of the United States, Johnson might, if the natural conditions were right and a careful selection made, achieve this result. A water hole might often be the only water for miles around. Small streams, in this semi-arid region, had the habit of disappearing underground to reappear for short stretches further down. By selecting a homestead with these natural features in mind and by using his rights under the Preemption, Desert Land, and Timber Culture Acts, Johnson, by merely enclosing his own property might control as effectively the ranges beyond, as if he had fenced those also.

Such a location as this is well described by a Colorado cattleman in his testimony before the Public Land Commission of 1879:

Wherever there is any water there is a ranch. On my own ranch [320 acres] I have two miles of running water; that accounts for my ranch being where it is. The next water from me in one direction is twenty-three miles; now no man can have a ranch between these two places. I have control of the grass, the same as though I owned it. . . . . Six miles east of me, there is another ranch, for there is water at that place. . . . . Water accounts for nine-tenths of the population in the West on ranches.[22]

[22] *Preliminary Report of the Public Land Commission*, 1879-1880, p. 297.

The presence of others along a stream, too long to permit individual control, meant that the exclusion of outsiders must come through some sort of an understanding among those already on the ground. Cooperation among neighbors in the conduct of their business resulted in the growth of a certain amount of range privilege and good will. Participation in the roundup, in the use of the common corrals, in the group protection against Indians, thieves, and predatory animals and, in some cases, in the group drive of the beef turnoff to the railroad could be permitted or denied to the newcomer. To deny such privileges, often appeared to be the only way of preventing the overcrowding of a range already taxed to its full carrying capacity. The success of such a method would, of course, depend upon the size of the outfit so denied and upon the amount of cooperation among the older stock growers.

The Montana stockmen, through their local associations, endeavored to control the range by this method. Notices such as the following appeared in the local newspapers warning off newcomers:

NOTICE TO STOCK GROWERS

At a meeting of the stockmen, owners of stock on the Musselshell, said range being defined as follows: to wit, Beginning at the mouth of Box Elder Creek, on the Musselshell River, thence up the Box Elder and Flat Willow to the head of the same, thence westerly along the divide to Judith Gap, thence westerly along the divide to Copperopolis, thence southerly along the divide to the divide between Fish Creek and Sweetgrass Creek, thence easterly along the divide between the waters of the Yellowstone and Musselshell rivers to a point opposite or south of the mouth of Box Elder Creek, thence north to Box Elder Creek. We, the undersigned stock growers of the above described range, hereby give notice that we positively decline allowing any outside party, or any party's herds upon the range, the use of our corrals nor will they be permitted to join in any roundup on said range from and after this date.[23]

This method of boycott, employed on the Montana ranges, was never very effective. In the first place, if a newcomer purchased cattle on an accustomed range, it was generally conceded that he bought with it the range privilege and good will. Once he was established, there was nothing to prevent him from purchasing additional cattle, thus jeopardizing all the other herds. Furthermore, it was soon discovered that if the stockmen of a certain area advertised their range by declaring it sufficiently stocked, it was evidence enough to the outsider that the former were in possession of a good thing and reaping huge profits therefrom. Newcomers might be excluded from the roundup, but in Montana there was nothing to prevent them from staging a roundup of their own, which meant that the cattle on the range might be worked over several times a season. Old-timers, who saw their cattle overworked, were glad to admit newcomers to the roundup to give their cattle a rest. Finally, as the cattle craze took hold, it was found that on many ranges it was not by any means the newcomers who were chiefly responsible for overcrowding, but the older outfits who were enlarging their herds by unrestricted speculation in eastern stock.

In Wyoming, the control that the cattlemen exercised over the grazing was immensely strengthened by the passage of the law of 1884, creating the legal roundup under the supervision of the Wyoming Stock Growers' Association. Until the passage of that law, the situation was not unlike that in Montana. In his annual report to the Secretary of the Interior, Governor Hale in 1883 described the difficulties that the outsider would have in getting on the Wyoming ranges, even before the law created any barriers:

---

[23] *Rocky Mountain Husbandman,* July 19, 1883.

There is at this stage of the progress of the industry a drawback facing a man of capital who wishes to begin stock raising by setting up a ranch and starting a new herd in Wyoming. Although it can hardly be said that the Territory is everywhere fully stocked with cattle, it is a fact that the more desirable ranges are generally occupied, and a man, taking into almost any locality a herd of 5,000 cattle, would meet with much difficulty in securing what are termed range rights from his neighbors. If a season of shortage in the grass crop were to follow, his annoyances would be considerable and the driving of his cattle over a large area of country in pursuit of food might be attended with loss.[24]

After the passage of the law of 1884, the range privileges and good will, without which the newcomer might be the victim of "annoyances," were vested in the Wyoming Stock Growers' Association. By the law of that year, no stock could be branded between February 15 and the commencement of the general spring roundup of the Association.[25] In practice, this meant that the calf roundup was wholly in the hands of the Association. Since the chief reason for rounding up in the spring was to brand the calves, and since any roundup before February 15 was dangerous to the cattle, the stock grower was practically prohibited from an independent roundup. True, he might stage a roundup while that of the Association was in progress or at some subsequent time, but this was impractical, if it were not actually prohibited by the other owners.

To all intents and purposes, the newcomer was forced to join in the regular spring roundup held by the Association and regulated by the law of the Territory. The advantages that would accrue from being a member in good standing of the Association were obvious. He would have a voice in the

[24] *Annual Report of the Governor of Wyoming to the Secretary of the Interior*, 1883 (Washington, 1883), 25.

[25] *Supra*, pp. 133-138.

conduct of the roundups and would receive all the protection which the organization offered.

If a range was beginning to get too crowded, the newcomer might find it difficult to obtain such admission. True, the law provided that any stock grower was eligible for admission, but the Association was judge of its own membership. Under its by-laws, the applicant for admission must have his name presented by a member. This application was reported on by an investigating committee of three. It was then voted upon, and three adverse votes were sufficient to reject it.[26] Rather than risk being denied the advantages that accrued from membership in the Association, the newcomer was likely to seek a less crowded area, where his chances for success were greater.

From the stockman's point of view, such measures to limit the number of operators in any given grazing area were justifiable acts of self-protection. To admit more cattle was to endanger the safety of all. The small operator trying to get a start and the granger edging out upon the ranges saw in such efforts an arrogant monopoly seeking to exclude men from the free pasture without the shadow of a claim. As they saw the herds of the leading members of the Association grow from year to year through constant purchase, it is not to be wondered at that they began to denounce "the cheek and unblushing effrontery," as one paper put it, "of those lords of creation, the cattle kings." [27]

These methods of preventing overcrowding need not have caused any fears that the "cattle kings" would succeed in monopolizing the ranges for any great length of time. They were all operating on a basis too insecure for that. The carrying capacity of the ranges was overtaxed in spite of all their efforts; indeed the worst offenders were members of the

---

[26] *By-Laws of the Wyoming Stock Growers' Association,* 7.
[27] *Bozeman Chronicle,* Sept. 12, 1883.

organization itself, who overbought and endangered their own safety. In sections where the cattlemen's organizations were weak, declarations against newcomers were disregarded, and new herds moved in, in spite of manifestos and warnings. And finally, the open range was being threatened by the appearance of two other groups for whom the threat of boycott had no terrors, the sheep herder and the granger. The first reduced the carrying capacity of the range, and the second narrowed down the area where the range cattle business could operate.

Sheep from California, Utah, and Oregon arrived on the northern ranges in the early seventies.[28] In Montana, they followed the extending cattle frontier eastward out into the open country. By 1881 the number of sheep assessed in the Territory exceeded the number of cattle. Meagher County, which was the leading cattle county in that year, was the leading sheep county as well. Although in Wyoming, cattle were well in excess of sheep during the range period, the western counties of Uinta, Fremont, and Sweetwater were predominately sheep counties by 1884. Against the sheep herder, fences or force were the only successful protective measures.

As for the granger, who began to arrive in eastern Wyoming just when the cattle boom was beginning to deflate, the law was all on his side. For him, the open range, where range rights and range customs held sway, did not exist; it was merely the public domain where a poor man might have a quarter-section of land and a home. The clash between this traditional figure of the frontier and the cattleman will be discussed in a succeeding chapter.

Try as he would, the cattleman, through his organizations, found no satisfactory solution for this problem of range con-

[28] Sheep appear on the 1866 assessment lists in Montana and on the 1870 lists in Wyoming. See page 230 for chart showing number of sheep in Wyoming, 1886-1905.

trol and was forced, in the last analysis, to fall back upon his own individual efforts to protect his interests. Where threats of boycott failed, a barbed-wire fence might succeed. Newcomers began to find long strings of fence reaching out across the range with an occasional placard, foretelling, in no uncertain terms, what would happen to him who dared cut the wire.

It was very natural for the settler along a stream, after he had fenced in his own tract, to run a line out upon his accustomed range beyond. We have noted how this practice had developed in the seventies in western Montana.[29] Here, it had not been a measure to prevent overcrowding but merely an easier method of handling the comparatively small herds owned by the ranchers in that section of the country. In the mountains, where timber was ready at hand, fencing was a practical proposition. Out on the treeless plains, where the only wood was the clumps of cottonwood along the margins of the streams, the fencing of the accustomed range had to wait until the appearance of barbed wire. In the year 1874, patents for barbed wire and for a machine for making it were taken out by J. F. Glidden of De Kalb, Illinois. By 1880 barbed-wire factories in the United States were turning out forty thousand tons of this cheap fencing; by 1890 this output had tripled.[30]

In the early eighties the practice of enclosing portions of the public domain with this cheap and easy fencing spread

[29] *Supra*, pp. 56-57. In describing the development of fencing in Wyoming, Governor Hale in his annual report to the Secretary of the Interior, 1883 (p. 49), said, "In building a fence to enclose his tract, the settler was induced to run it out upon the plains as far as the middle of the uplands, dividing the stream upon which he had settled from the one running next to it — in some places a distance of miles. He reasoned that no other settler could wish to take up the waterless highlands he thereby enclosed, as a range for his stock, and that in time the Government might afford him some lawful means of gaining possession of it."

[30] There had been earlier patents for barbed wire in 1867, but it was not

so rapidly that the whole range industry was in danger of being strangled to death in a web of its own making. Trail herds, on their way to the railroads or to distant ranges, found long drift fences barring their path. Long-established routes of travel were blocked and mail carriers complained that they were forced to detour around some cattleman's enclosure. A Texas governor was forced to ask the state legislature to free the county seat of Jones County, which was completely circumscribed by a fence, fifteen miles distant, having but two gates.[31] Goodnight, at the head of the Red River, had up two hundred and fifty miles of fence in 1884.[32] He and his neighbors were running a single line of fence from the western border of the Indian Territory straight across the Pan Handle and on into New Mexico for thirty-five miles. It was hoped that this would keep the Kansas herds from drifting south upon the Texas ranges before the winter storms.[33] In Colorado, the Arkansas Cattle Company had forty townships under fence by 1884, an area of almost a million acres.[34] Small outfits, particularly in western Kansas and in the Indian Territory, combined into huge "pools," chiefly for fencing purposes, and fenced off hundreds of thousands of acres.[35]

until 1874 that a practical method of manufacturing was developed. The production 1874-1907 is given as follows (*Enc. Brit.*, 13th ed., III, 384-385):

Year	Tons	Year	Tons	Year	Tons
1874	5	1878	13,000	1900	200,000
1875	300	1879	25,000	1907	250,000
1876	1,500	1880	40,000		
1877	7,000	1890	125,000		

[31] *Drover's Journal* (San Antonio) quoted in the *Rocky Mountain Husbandman*, Jan. 31, 1884.

[32] *Breeder's Gazette*, V, 674, May 1, 1884.

[33] *Breeder's Gazette*, V, 50, Jan. 10, 1884.

[34] Report of the Commissioner of the General Land Office on the Unauthorized Fencing of the Public Lands, *Sen. Ex. Doc.* No. 127, 48 Cong., Sess. 1, 1883-1884, p. 2.

[35] E. E. Dale, "History of the Ranch Cattle Industry in Oklahoma" in the

On the northern ranges, cattlemen, large and small, were buying wire. In the report of the Land Commissioner on illegal fencing, the older cattle counties in both Montana and Wyoming were listed as containing illegal enclosures of the public domain. Neither the eastern counties of Montana nor the northern counties of Wyoming were cited. Apparently these sections had not yet felt the pressure that had resulted in extensive fencing in the older areas.[36] Laramie County, Wyoming, appeared to contain the worst offenders on the northern ranges, for in 1886, ten large companies were listed as having had illegal enclosures, among which was the Swan Land and Cattle Company with one hundred and thirty miles of illegal fence.[37]

Complaints and petitions for relief began to come into the Land Office. Cattlemen were accused of setting up small barbed-wire kingdoms to the great detriment of the sturdy pioneer looking for land. The investigation conducted by the Land Office piled up a mountain of testimony as to the high-handed practices out on the plains. Settlers were counseled by the Secretary of the Interior in 1883 to cut all fences barring them from the land on which they desired to settle.[38]

The din raised over this evidence of the cattleman's greed finally moved Congress. In 1885 a law was passed designed to expedite prosecution of those who stretched fences out

Annual Report of the American Historical Association, 1920, pp. 317-319. This article gives an account of the largest of the pools of the Southwest, the Cherokee Strip Live Stock Association. Other pools of a similar nature in Kansas were the Walnut Valley, Smoky Hill, Eagle Chief, and Comanche pools. Notices of these and their activities are to be found in the local news prints.

[36] Unauthorized Fencing, *op. cit.,* 19-24.

[37] *Annual Report of the Commissioner of the General Land Office,* 1886, p. 464.

[38] Letter of Secretary Teller to Commissioner of Agriculture, quoted in *Rocky Mountain Husbandman,* April 12, 1883.

upon the public domain.[39] Under this law, suits were instituted in the range country, and the illegal fences began to come down. But they never entirely disappeared. In 1904-1905 another Public Land Commission of investigation reported that illegal fences were still to be found in many sections. In commenting upon this, it said, "Where such illegal fencing of the range is mutually agreed upon, there is no complaint made against anyone and usually the Government meets with limited success in securing convictions." [40]

As a matter of fact, the stringing of hundreds of miles of fence across the open range was far more dangerous to the property of the cattlemen themselves than it was to the rights of the settler. In the first place, by limiting the free movement of the herds, it prevented even grazing, and in the second, it increased the winter losses. Cattle drifted for miles before a blizzard and if they could keep moving, tails to the wind, they would probably survive. Once they came to a halt before an impassable barrier, they were lost, unless the storm abated. In commenting on the fencing craze in Texas in 1883, an editor wrote the following prophecy: "Under the old regime, there was a loose adaptability to the margins of the ranges where now there is a clear-cut line which admits of no argument, and an overstocked range must bleed when the blizzards sit in judgment." [41] Two years later, the carcasses of thousands of cattle along the fence lines of western Kansas, Colorado, and Texas were the price that the cattlemen themselves paid for fencing the open range.

The enclosure of the accustomed range was therefore neither practical nor legal. The purchase or lease of the range might be made legally possible by an alteration of the

[39] 23 *U. S. Stats.* 321.

[40] Report of Public Land Commission, 1905, *Sen. Doc.* No. 189, 58 Cong., Sess. 3, Appendix p. 10.

[41] *Mobeetie* (Texas) *Panhandle*, quoted in the *Breeder's Gazette*, IV, 526, Oct. 18, 1883.

existing land laws to fit the range country. But that even such a change would result in a practical solution of the cattle-men's difficulties was open to question.

By the middle of the seventies, it was clear enough that the laws under which title could be obtained to any portion of the public domain did not fit the environment west of the hundredth meridian. As early as 1875 the Commissioner of the General Land Office in his annual report was urging Congress to modify the existing laws to meet the needs of the settlers in the arid West.[42] Two years later, the Secretary of the Interior called attention to the fact that it was not the farmer but the stock-growing industry that was spreading out over these areas and utilizing the land, not only without the authority, but without the protection, of the law. The small subdivisions, suitable for agricultural settlement, provided for under the Homestead and Pre-emption Laws, were inadequate in a very considerable part of the remaining public domain.[43]

Absurd as it was to talk about one-hundred-sixty-acre homes for poor men in a country where it took anywhere from ten to thirty acres to furnish grass enough for a range steer, the country in general continued to think of this prob-lem of adapting the land laws to the arid West in terms of agriculture as it was known in the Middle West. Congress was not interested in the cattlemen and the methods they were evolving for the utilization of these regions. As on the older frontiers, they were regarded as merely an advanced screen ahead of the real conquerors of the land, the pioneer farmers. If the arid West did not welcome the farmer as the rich prairie soil had done, it might be induced to do so by congressional legislation. The thing to do was to adapt

[42] *Annual Report of the Commissioner of the General Land Office*, 1875, pp. 6-7.

[43] *Annual Report of the Secretary of the Interior*, 1877, p. 20.

the High Plains to the farmer and not the farmer to the High Plains. The Land Office and, through it, Congress was made to believe that if the western farmer were wheedled into planting enough trees, rainfall would be increased to such an extent that the arid West would succumb to the all-conquering farmers, as had the buffalo and the Indian. In 1873, Congress was persuaded to pass the Timber Culture Act, which gave a quarter-section of land in addition to the homestead quarter-section to the pioneer who would plant forty acres of it with trees. Rain-making by legislative fiat was something new, even in the varied history of our public land legislation.[44]

Irrigation had long been practiced in the West, and wherever there were favorable locations for easily constructed dams, small irrigation projects had been developed, chiefly through individual effort. Although agriculture by irrigation requires a highly specialized technique, which was a closed book to the eastern farmer, Congress was willing to legislate irrigation ditches into existence also. If 160 acres of sagebrush were not enough to get the farmer started, he might be induced to follow the western star of empire by offering him 640 acres more. So, in 1877, the first Desert Land Act was passed, which, it was hoped, would lure the farmer by permitting him to buy 640 acres at a dollar and a quarter an acre, twenty-five cents to be paid at the time of entry, and the balance when the final proof was made. The only other condition attached was that the farmer should irrigate the whole section of 640 acres in three years.[45] This was expecting a miracle, second only to the rain-making act of 1873. The tradition of small farm homes for poor men lived on in land legislation and political oratory long after it ceased to be a practical proposition on the American frontier.

[44] 17 *U. S. Stats.* 605.   [45] 19 *U. S. Stats.* 377.

While Congress was legislating the farmer westward, there were attempts in other quarters to deal realistically with the problem of the High Plains. In 1878 the report of Major James W. Powell on the *Lands of the Arid Regions of the United States* was submitted to the Secretary of the Interior, who on April third transmitted it to Congress.[46] The experience that this geologist had had as head of several government exploring expeditions in the arid West gave weight to his report.

The Powell report contained three important proposals: first, the classification of the lands of the public domain into mineral, timber, coal, irrigable, and pasturage land; second, a change in the system of survey; and third, a modification of the homestead system to fit the new environment. The first proposal recognized the existence of vast areas of western country which were useful only for pasturage and urged that they should be classified as such. The second proposal was based on the argument that the all-important topographical feature of the arid West was water, and survey lines should be made to conform to it. It was absurd, urged Powell, to waste time and money laying off rectangular sections, thousands of which would be of no practical value as independent units. Survey lines should be so drawn that the greatest number of water frontages could be obtained.

The third proposal was that Congress should provide for the two groups who alone would be able to utilize the arid West, the irrigators and the cattlemen. Powell recognized the crucial part played by cooperation in the Far West and provided for it. He proposed that the lands suitable for irrigation should be divided into irrigation districts, inhab-

[46] Maj. James W. Powell, *Report on the Lands of the Arid Regions of the United States* (Washington, 1878). Found also as *House Ex. Doc.* No. 73, 45 Cong., Sess. 1 and 2, 1878.

ited by not less than nine irrigators, who would receive title to eighty acres apiece, upon the completion of a common irrigation project. For the herdsmen, he provided for similar pastoral districts, organized by nine or more persons, who, upon organization, were to receive 2,560 acres of pasture apiece, with access to water. These proposals were embodied in two bills, which were transmitted to Congress with the report.

It was too much to expect Congress to act upon any such revolutionary scheme, which would upset the whole system of survey and disposal then in operation. The Powell Report did have its effect, however; it started a movement toward adequate land classification. In the law approved March 4, 1879, creating the United States Geological Survey, the director of that work was instructed to make such a classification.[47]

In addition to this, Congress was sufficiently aroused to create by the same act a Commission on the Public Lands. This Commission was instructed to codify the existing land laws, report on a possible change in the present system of survey, and make recommendations as to the best methods of disposing of the land of the United States in the West to actual settlers. After spending the summer of 1879 in all the western states and territories interviewing farmers and stockmen, the Commission made its report.[48]

At the hearings held in the western country, the Commission tried to get at the desires of the western stock grower and agriculturist. Was he in favor of a change in

---

[47] 20 *U. S. Stats.* 394. The legislative background described above is ably presented by B. H. Hibbard in his *History of the Public Land Policies* (New York, 1924), 411-455, 496-501.

[48] The preliminary report was submitted to the second session of the 46th Congress. (*House Ex. Doc.* No. 46, 46 Cong., Sess. 2, 1879-1880.) The report contains the testimony taken in the field. The final report was submitted at the next session. *House Ex. Doc.* No. 47, 46 Cong., Sess. 3, 1880-1881.

the land survey? Did he favor the pastoral homestead of Powell? Would a gradual reduction in the price of the unsold land meet the problem? Was he in favor of some sort of leasing system?

The answers to these questions were varied, for the cattle country was not of one mind as to the possible benefits that such changes would bring. Although the cattleman as an individual and in his organizations was anxious to prevent overstocking by excluding newcomers, he was aware of the fact that a restraint on them, imposed through a change in the land laws, might be a restraint upon himself. Many were more willing to take their chances on the open range than they were to reduce their present profits by paying for the exclusive use of a part of the range by lease or by purchase at a reduced figure. Private control meant the erection of fences, and as we have noted, many cattlemen, large as well as small, saw the danger such barriers presented to the industry as it was then being operated. Only the largest operator would be able to lease or buy enough land to escape this danger, and even he would prefer to have his cattle out on the unfenced range when the storms came.

The chief objection, however, was on the ground that such proposals would create vast monopolies. "Beat the land grab" was the most definite answer that the Commission received. It was a poor privilege that the neglected cowboy, "who had subdued the wilderness and laid the foundations of civilization was asking from the Government, the privilege of allowing his cattle to feed unmolested on this empire of grass." [49] The Montana legislature in 1881 memorialized Congress to the effect that such changes "would be inimical to the true interests of Montana . . . . and would place large tracts . . . . into the hands and under

[49] *Rocky Mountain Husbandman,* Oct. 30, 1879.

the control of monopolists and speculators." [50] Eastern land
sharks were blamed for this "gigantic steal." The Montana
delegate in Congress asserted that these proposals "were
born of the desires of great cattle kings and wealthy stock
companies." [51] In the annual meeting of the Wyoming Stock
Growers' Association, which might be supposed to contain
some of the "cattle kings," if any such existed on the northern ranges, strong resolutions were unanimously passed on
the identical grounds that such a system would concentrate
the business in the hands of a few men, that it was undemo-
cratic, and opposed to our theory of government.[52]

The Commission, in its final report, urged a system of
classification, a gradual reduction of the price of unsold
lands, and the adoption of the Powell pastoral homestead.
It nullified this proposal by adding, "There is a deep-seated
conviction in the minds of the majority of the people of this
country that a system which tends toward monopoly or even
permits the aggregation of very large tracts of land into the
ownership of a single person is unjust." [53] After such a warning, Congress, always sensitive to popular sentiment against
land monopoly, could be counted on to oppose any tinkering
with the land laws along the lines laid down by the Commission.

In 1879, the range cattleman was still a frontier figure,
operating with little or no capital outside of what he had
invested in his herd. For him, the open range was a practical proposition, for it constituted a satisfactory basis for
his prosperity. He did not have the money to buy the land
that supported his herd, even though the price was placed at

[50] *Laws of Montana Territory,* 1881, Sess. 12, pp. 130-131.
[51] *Cong. Record,* 46 Cong., Sess. 2, pp. 3925-26.
[52] Minute Book, Wyoming Stock Growers' Association, Nov. 18, 1879, pp. 37-45.
[53] Preliminary Report, *op. cit.,* pp. v-viii.

a minimum, nor did he see any reason why he should buy. If his range became depleted, he followed the usual frontier method and moved on.

The discussion of the proposals of the Land Commission of 1879 in the annual meeting of the Wyoming Stock Growers' Association reflected very clearly this attitude and is an index of what the range-cattle industry thought of itself and of its future just on the eve of the boom period. Among the reasons given for opposing the contemplated changes was the belief that the future prosperity of Wyoming might come through farming and that the selling of large bodies of land might prevent settlement.[54] That the cattleman was prepared to keep ahead of this settlement, unhindered by ownership of depleted ranges, is clear from the following quotation from the resolutions adopted by the Association:

[Resolved] That in our opinion the question of whether grass will not disappear from the ranges with constant feeding is still unsettled, and that the stock business will not warrant the investment of so large a per cent of capital as one-sixth in what may, in a few years, be barrren and worthless property.[55]

As to the question of overcrowding, the Wyoming Association felt no uneasiness on that score. "Thus far," the resolution states,

self-interest has proved a safeguard against heavy stocking of the range, and . . . . we would rather trust for maintaining our rights, free from encroachment, to the community of interests and the sense of equity that rule here than to see a system adopted that must excite serious quarrels between occupants of ranges and a bitter feeling of injustice among the best stockmen, who, unable to buy, ... will see themselves deprived of their business.[56]

[54] Minute Book, Wyoming Stock Growers' Association, 38.
[55] *Ibid.,* 39.
[56] *Ibid.,* 40.

This was in 1879. Four years later, as we have noted, lines of fence were going up all over the range and the "land grab," against which the cattlemen had cried out, appeared to be under way. The Land Office reported that not only were vast tracts of public domain being fenced off without a shadow of a claim, but the most valuable lands of all, those upon which there was water, were passing into private ownership through the practice of the grossest kind of fraud and collusion.[57] The chief offenders in all this were the cattlemen who had opposed the proposals of the Land Commission of 1879 on the ground that such changes would foster land monopoly.[58]

The proposal to provide for some system of leasing the range, which had been generally turned down in 1879, was now put forward, not by a government commission, but by the cattlemen themselves. At the meeting of the National Cattle Growers' Association in St. Louis in 1884, a resolution was introduced, urging Congress to "enact such laws as will enable the cattlemen of the West to acquire by lease the right to graze upon unoccupied lands." [59] This resolution received the ardent support of the Wyoming delegates, the leaders of the delegation speaking to the resolution, urging that the cattle industry must be given a chance to obtain some sort of control over the lands upon which it operated. These same gentlemen, Carey, Sturgis, and Babbitt, had urged their Association in 1879 to reject all the proposals of the Land Commission, one of which provided for a leasing system.

Much had happened on the northern ranges in the three or four years intervening to absolve these men from the charge of inconsistency. The sudden influx of millions of

---

[57] *Annual Report of the General Land Office*, 1885, pp. 48-67.

[58] *Ibid.*, 1886, pp. 50-69.

[59] *Proceedings of the National Cattle Growers' Convention* (St. Louis, 1884), 81-85; Clay, 117-118.

capital between 1879-1884 had changed the whole character of the industry and with it the attitude of the cattlemen on the land question. The ranges had suddenly become crowded with cattle, purchased at high prices. Some of these cattle were of a very distinctly superior breed to the older Texas herds, and represented an investment of such proportions that every means possible must be taken to afford them adequate protection.

The erection of illegal enclosures was, as we have noted, a method of securing to the cattleman the exclusive control of that part of the public domain which he had come to regard as his accustomed range. The attempt to prevent overcrowding by this method had involved so great a danger to all, that the action of the Government against such illegal fencing was welcomed in many quarters by large and small range operators alike.

Another process of far more importance to the future of the range was under way. Forage, other than the range grass, was becoming increasingly important, particularly to those with valuable herds. When the early cattleman had come to a patch of wild hay along some stream, he had allowed his cattle to graze it off along with the less luxuriant growth on the benches beyond. A few tons might be put up by the cowboys for the riding stock, and in some favorably located spots hay ranches had developed to furnish the motive power in stage coach and freighting days and to supply the army posts. Beyond this, little advantage had been taken of this natural resource. The cattle boom changed all this. There was a mad scramble for ranch properties. Speculation in ranch sites became as hectic as speculation in herds. Large companies began to divert some of their capital from the buying of cattle to the purchase of land. This shift of the capital basis of the industry from cattle, to land and cattle, resulted in a tremendous increase in the disposal of land by the Government under the various land laws and the appearance of land frauds far more serious than those of illegal fencing.

Fraud in the disposal of the public domain is no new thing in the history of our public land policy. It never reached larger proportions, nor developed a greater wealth of ingenuity in the methods employed, than during the last half of the eighties. The alarm that such methods aroused was equally great. The people of the United States had become conscious that there were limits to our national estate and that the remaining areas were being all too rapidly dissipated. The farmer in the Dakotas, Nebraska, and Kansas had made contact all along the line with the cattle frontier. Here he found great land and cattle companies, many of them financed by foreign capital, brazenly stealing the land of the United States to the injury of all honest farmers. It made little difference that much of the land was of a marginal character, which would never be utilized for agriculture; the farmer wanted to retain his right to fail as well as to succeed on a quarter-section of free land. The agrarian discontent, which was expressing itself in attacks on the railroads, the "grasping" corporations and the "money power" of Wall Street, was also turned upon the cattle country. The arrival of the Democrats in Washington in 1885 as the party of reform resulted in an overhauling of the Land Office and a general airing of the abuses that had been permitted under the previous Republican administrations.

William Andrew Jackson Sparks, who was appointed by President Cleveland to the post of Commissioner of the General Land Office, more than fulfilled Democratic expectations in exposing the laxity in the administration of the public domain under the Republicans. He set to work with all the zeal of the true reformer, a zeal that later proved to be embarrassing enough to his superiors to result in his removal from office. In his first annual report, 1885, he had much to say about the cattlemen. "In many sections," he declared, "notably through those regions dominated by the cattle-raising interests . . . . examinations wherever made had developed at all points that entries were fictitious and fraudulent and made in bulk through concerted methods adopted by organizations that had parcelled out the country among themselves." From all these sections where examination had been made came "one common story of widespread, persistent land robbery committed under the guise of the various forms of public entry." [60]

The lands that the cattlemen were getting hold of were those along the streams where there were natural hay lands or where there was a possibility of hay crops with a minimum of irrigation. The more valuable the herd, the greater the pressure on its owners to secure such lands. Writing in 1883, Governor Hale cited as an example of this progress in cattle raising the Wyoming Hereford Association, a company devoted to the raising of fine stock.

The pasture lands [he wrote] owned by this company border Crow Creek (six miles east of Cheyenne) for a distance of several miles. About fifty tons of hay were cut the past season, but the cattle have since grazed in the meadows, the strong close-set grass growing rapidly. The meadows average perhaps half a mile in width, [with] high but broken bluffs to the north and south; the cattle finding shelter among them, the greater part of the winter. Nothing, however, is risked as to weather in the management of the herd of pure blood Hereford stock now grazing on the ranch. [61]

What was being done by this company was being done to some degree by every company that could get control of a favorable location. Homestead, preemption, and the desert

[60] *Annual Report of the Commissioner of the General Land Office*, 1885, p. 49.

[61] *Annual Report of the Governor of Wyoming to the Secretary of the Interior*, 1883 (Washington, 1883), 28.

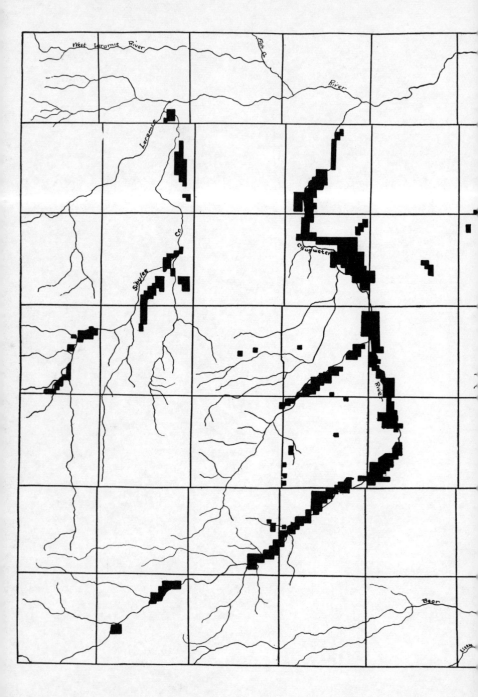

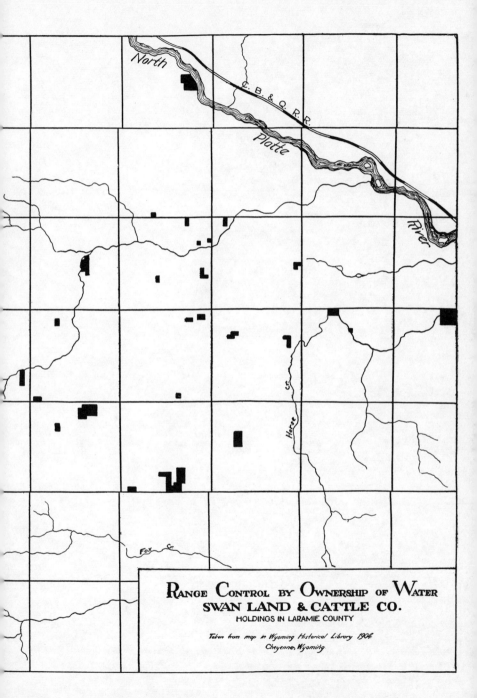

RANGE CONTROL BY OWNERSHIP OF WATER
SWAN LAND & CATTLE CO.
HOLDINGS IN LARAMIE COUNTY

Taken from map in Wyoming Historical Library 1904
Cheyenne, Wyoming

land laws were used to obtain these stream sites. Cowboys, employed by the company, were induced to use their rights to enter lands under the preemption and desert land acts, the company paying them for so doing. Wherever the land was worth it, the cattlemen were willing to pay the initial twenty-five cents an acre for a desert section, which gave them exclusive control for three years at what amounted to a yearly rental of eight and a third cents an acre. At the end of three years, if the land was worth retaining, full title might be secured by paying the additional dollar an acre and demonstrating that the land had been reclaimed by irrigation ditches thereon. A few furrows run out in any direction with no regard for the location of water nor the contour of the land were enough to satisfy the local land office.[62] As early as 1880, the Commissioner had been compelled to issue a circular, cautioning the local officers that land upon which wild hay grew was not to be regarded as desert land.[63] But as no classification had been made, the discretion lay with the local office, and the frauds under that ill-advised and impossible act continued.[64]

[62] In addition to the material on the subject of fraud in the acquisition of land in the West found in the annual reports of the Commissioner of the General Land Office, the following special reports contain a mass of testimony: *Sen. Ex. Doc.* No. 127, 48 Cong., Sess. 1, 1884, on unauthorized fencing of the Public Lands; *Sen. Ex. Doc.* No. 225, 49 Cong., Sess. 1, 1886, on land entries canceled for fraud; *House Ex. Doc.* No. 232, 50 Cong., Sess. 1, 1888, on the use of public lands by graziers. Two reports of the House Committee on Public Lands contain some additional information: *House Report* No. 1834, 47 Cong., Sess. 2, 1882, and *House Report* No. 1325, 48 Cong., Sess. 1, 1884.

[63] *Annual Report of the General Land Office*, 1880, p. 87.

[64] The map showing the land holdings of the Swan Land and Cattle Company in Laramie County, Wyoming illustrates the methods used by the cattleman to control the range by controlling the water. By owning the land along the Chugwater and the Sybylee, the company controlled not only those valleys but the high land between them. Between the Chugwater and the North Platte, the holdings were more scattered, consisting of waterholes between the watercourses, corrals, and holding grounds.

The stiff regulations issued by Sparks to prevent the continuation of such frauds and the wholesale cancellation of entries suspected of fraud brought forth a storm of protest from the cattle country. Governor Warren of Wyoming, himself a large stockman, declared,

. . . . that land matters in Wyoming are misunderstood and misjudged [and that] . . . . if an over-zealous course is pursued and the acquirement of land by bonafide entrymen is made so difficult as to amount to almost proscription, very great injury is done to the class sought to be benefited by such efforts. . . . . Well meant, iron-clad instructions do not so much hinder frauds as they embarrass and impoverish the poor pioneer.[65]

In spite of Governor Warren's tender solicitude for the poor pioneer, the cattlemen were interested in the real pioneer industry, their own. The time had come when their continued prosperity depended on the acquisition of as much land as possible. They no longer thought of themselves as temporary occupants, permitted to utilize these acres until the farmer arrived, but as natural, permanent, and all important. "Considered as between the farmer and the stockman," wrote Sturgis, the secretary of the Wyoming Association in 1884,

the stockman has on his side, the following arguments which it would be hard to refute. We occupy a country in which irrigation is a conceeded necessity to agriculture. Nine-tenths of the surface is totally unfit for the plow, even under the most favorable conditions of capital. The stockmen utilize it all. Displace him and his capital and fifty years must pass before these plains can be made to produce the same taxable wealth in any other form.[66]

Forty-one years later, in 1925, the accuracy of Sturgis's estimate is the best support of the cattleman's argument. Of

[65] *Annual Report of the Governor of Wyoming to the Secretary of the Interior*, 1886 (Washington, 1886), 5.
[66] *Cheyenne Daily Sun*, April 4, 1884.

the sixty-two and one-half million acres in the state, about one-third were privately owned. Only 8.8 per cent of this land in private ownership was devoted to agriculture, the remainder was in privately owned pastures. Thus, less than 3 per cent of the total area of the state had become agricultural land forty-one years after the Sturgis report. The total value of farm crops in 1925 was $31,509,000. In 1884 the estimated value of livestock was $23,606,000.[67]

Land laws, made wholly for the farmer who was expected to displace the range cattleman, invited fraud. The former found them impossible of honest application, and the latter regarded them as the product of middle-western and eastern ignorance. Congress, prodded into action by the President and the Secretary of the Interior, found that it had no ideas on the subject. Members, with little or no knowledge of the conditions as they actually existed west of the hundredth meridian, took their cue from the fulminations of Sparks and declaimed against the "land grabbers" and the "cattle kings." Delegate Carey, listening to a denunciatory speech from an Indiana member in Congress, resorted to the *tu quoque* argument by declaring that more land frauds had been committed under the Swamp Land Act in the State of Indiana than in the whole arid West.

Sparks became the object of attack and ridicule in the cow country. "Thou shalt have no other gods," chanted the *Cheyenne Sun*, "than William Andrew Jackson Sparks, and none other shalt thou worship. Thou shalt not raise cattle upon the land, neither sheep or asses nor any living thing, but only corn the same as in the State of Illinois." [68] His resignation in 1887, due to an excess of zeal, was a cause for general rejoicing in the West.

---

[67] *Wyoming Agricultural Statistics,* Joint Bulletin of the United States Department of Agriculture and the Wyoming State Department of Agriculture (Cheyenne, 1925), 4-11.

[68] *Cheyenne Daily Sun,* Feb. 27, 1887.

In 1888 Congress did make a feeble effort to prevent the further alienation of those parts of the public estate where agriculture by irrigation was at all possible. All lands suitable for irrigation projects and all lands susceptible of irrigation were withdrawn from entry. If at some future time the conditions warranted it, these lands could be opened by the President to settlement.[69] The great difficulty was that the geological survey had not been made and in most cases was not made for several years, so that there was no data available on which to select the lands to be withdrawn.

The Desert Land Act, under which most of the fraudulent practices had grown up, was changed in 1890, by reducing the size of the entry from 640 acres to 320 acres.[70] The absurd requirement of the earlier act that the entryman must irrigate the whole section in three years was modified the next year by stipulating that only eighty acres should be under ditch at the end of the three-year period. Specific requirements as to the value of the improvements were also laid down.[71] That these changes did little to adjust the land policy of the United States to the High Plains was shown by the fact that fifteen years later, another land commission, appointed by President Roosevelt in 1903, found the same frauds obtaining in the administration of the Desert Land Laws.[72]

In his efforts to find a practical land basis for his changing industry, the cattleman turned to the purchase of railroad land. The transcontinental roads were among the first to become aware of the importance of the cattle industry on the High Plains, its capacity for expansion, and its relation to the land. In 1875 the government directors of the Union

[69] 25 *U. S. Stats.* 527.

[70] 26 *U. S. Stats.* 391.

[71] *Ibid.,* 1096.

[72] Report of the Public Land Commission, 1905, *Sen. Doc.* No. 189, 58 Cong., Sess. 3.

Pacific pointed out that the herds on the plains of the Platte were constantly increasing and that the "aggregate number of cattle was becoming fabulous. These plains," the report continued, "furnish an unsurpassed grazing range and the lands belonging to the government and to the company ought to be placed under some well devised system of pasturage . . . . whereby the growth of cattle may be fostered and the lands made immediately remunerative." [73] In 1877 the directors were complaining that none of these lands west of the hundredth meridian were being sold. There was no inducement for a grazier to purchase from the railroad an alternate section that would be surrounded on all sides by government land. If the Government could be induced to lease its land, then the railroad would be able to lease also. By such a joint system, blocks of fifty to five hundred square miles could be leased "at such a rental and for such a term of years and with such restrictions as will best protect the Government and the railroad company." [74]

Such a discussion was little better than an academic one in the seventies, for the cattleman of that period was not interested in schemes whereby he might be allowed to lease land in "blocks of fifty to five hundred square miles." There was plenty of room; he grazed his cattle where he pleased on railroad or government land alike. There was no reason why he should urge the leasing of land that was costing him nothing to use.

Five years later, section lines meant something more to him than markings on the map. The system of alternate railroad sections, modeled like the land laws, for middle-western farmers, was as awkward for the cattleman seeking private range, as it was for the railroad trying to sell land. "The

[73] Annual Report of the Government Directors of the Union Pacific Railroad, 1875, *Sen. Ex. Doc.* No. 69, 49 Cong., Sess. 1, pp. 99-100.
[74] Report, 1877, *op. cit.*, 134-135.

present inclination of the cattle interests is toward ownership of the range enclosure by fencing of the same and the
better development of stock," the Union Pacific directors
reported in 1883, "all of which is, if not rendered impracticable, greatly embarrassed by the existence of alternat-
ing government sections." [75] To remedy this, several plans
were suggested by Land Commissioner Burnham of the
Union Pacific. The first proposal was to exchange all sections of railroad land on one side of the track for all sections
of government land on the other, thus making a solid body
of land twenty miles wide on one side of the track belonging
to the railroad, which might be leased or sold to the cattlemen. The second proposal was a modification of the land
laws so as to allow for a sale of grazing lands in larger
tracts, government and railroad sections alike, at a reasonable price. The third plan was a long-time lease of the government sections for grazing purposes.[76] Action on any of
these proposals was impossible, for it would immediately
raise the cry of land monopoly.

In 1884 the land sales of the railroads in the cattle country began to mount. In 1883 the directors of the Union
Pacific had reported that of the 4,762,174 acres of land
along its route in Nebraska, 2,580,000 acres remained unsold, practically all of its estate west of the hundredth
meridian. In Wyoming of the 4,582,520 acres only 2,520
acres had been sold. This land was practically all grazing
land with an average value of one dollar per acre.[77] But in

---

[75] Report, 1883, *op. cit.*, 213.   [76] *Ibid.*, 214.
[77] *Ibid.*, 211-212. The figures are given as follows:

State or Territory	Original Grant	Am't Unsold, 1883
Nebraska	4,762,174	2,580,000
Wyoming	4,582,520	4,580,000
Utah	1,107,520	1,027,000
Colorado	688,900	690,000
Total	11,141,114	8,877,000

1884 the company began to dispose of these arid sections.

After consideration for several years of the most equitable and satisfactory manner of dealing with the lands of the company in Wyoming, they were, for the first time, in the spring of 1884, sold in large, compact tracts with reference to their pre-occupation, so far as they were occupied, for grazing and ranch purposes. A prompt and ready sale during 1884 disposed of the great bulk of available land, the transactions aggregating 2,081,130 acres and representing an almost solid, continuous body from the eastern boundary of the Territory, west to the vicinity of the North Platte River at Ft. Steele. . . . . While the prices have, as compared with Nebraska lands, been average ones, they have been desirable ones for the company, and one apparent result of these sales has been the introduction of foreign capital, the strengthening of that already invested, and the development of natural resources which will make the lands more valuable.[78]

The arrival of the Northern Pacific on the Yellowstone, a strong competitor for the business of the northern cattle range, may have influenced this decision. Railroad land, sold on favorable terms could be used to build up business along the line and injure a road with a less liberal land policy. As early as the summer of 1882, the Northern Pacific was selling alternate sections of grazing lands. One company, financed by eastern and foreign capital, bought 750,000 acres along the Little Missouri and Powder rivers at a dollar an acre.[79] The Union Pacific, in order to hold the cattle outfits on the older ranges and nearer its line than that of its rival, was forced to sell its land on equally favorable terms.

[78] Letter, Leavitt Burnham, land commissioner of the Union Pacific, to T. L. Kimball, General Traffic Manager, in the *Annual Report of Governor Warren to the Secretary of the Interior*, 1886 (Washington, 1886). The Swan Land and Cattle Company was one of the largest purchasers, buying up all the alternate sections for a distance of nearly fifty miles along the route. (*Breeder's Gazette*, VI, 21, Aug. 24, 1884.) Clay (p. 202) put the amount at 555,890.27 acres.

[79] *Helena Weekly Herald*, Aug. 17, 1882.

This purchase of alternate sections of railroad land appeared to many companies to be a solution of their land problem. Because of the checkerboard arrangement of the railroad sections, the purchaser of fifty sections of railroad lands, by fencing merely his outside sections, got control of some thirty or forty alternate government sections. This meant that the cattleman was getting practically exclusive pasturage for a little better than fifty cents an acre. Since the terms offered by the railroads were very easy, the Union Pacific allowing payment in annual installments over a period of ten years with 6 per cent interest, exclusive range could be obtained for a little better than five cents per acre per year.

On such terms as these, many companies thought they could operate. In place of the bold fencing of any part of the public domain, which had been prosecuted under the Enclosures Act of 1885, there appeared on the railroad grants new barbed-wire principalities. Fences were run along the purchased railroad section, four or five inches from the section line, beyond which lay a government section for which the fencer had no title. When the fence came to the section corner, it stopped a few inches from the line. Here was a geometrical point, where the corners of four sections came together, two belonging to the fencer and two to the Government. Obviously a fence could not be run across this geometrical point without encroaching on one or the other of the government sections. So the fencer stopped his fence there, went across to the next railroad section and started a new fence, four inches from the section line, leaving an eight-inch gap where one section ended and the other began. If, by this process of fencing his outside sections, the alternate sections of government land were enclosed, it appeared to the cattleman as a fortunate arrangement, incident on the

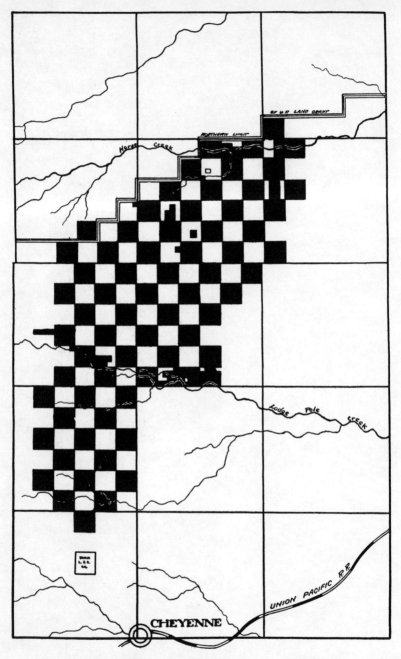

Labels within map: OF U.P. LAND GRANT, NORTHERN LIMIT, Horse Creek, Lodge Pole Creek, UNION PACIFIC R.R., CHEYENNE

RANGE CONTROL BY OWNERSHIP OF ALTERNATE
RAILROAD SECTIONS

John Arbuckle Ranch, Laramie County, Wyoming, from a map in the Wyoming
State Historical Library.

Government's policy of granting alternate sections to the railroads.[80]

Proceedings were instituted to enjoin the owner of railroad sections from constructing such fences, on the ground that a portion of the public domain was enclosed in violation of the act of 1885. The first suit of this kind was brought into the District Court of Albany County, Wyoming, in 1888 by the United States District Attorney. An injunction was asked against the Douglas, Willan-Sartoris Company, which, in the process of fencing in a tract of 81,440 acres in the manner above described, had enclosed over two hundred even-numbered townships, amounting to 38,720 acres, which belonged to the United States. The District Court, having dissolved the temporary injunction obtained by the Federal authorities, dismissed the petition. Whereupon the case was brought before the territorial Supreme Court on a writ of error.[81]

The Supreme Court of the Territory affirmed the action of the lower court, taking the ground that although the building of such a fence did have the effect of severing from the public domain some two hundred sections as the defendants admitted, this was incidental to the perfectly legal act of fencing. The plaintiff had held that such a fence interfered with the agents of the government and its beneficiaries from getting to their own land. So, said the court, would a fence made completely around each individual section, which could not possibly be regarded as illegal. Should the Government have a way over the defendant's land? Not unless such a way was paid for and along a prescribed line. True, a way of necessity might be granted to an actual occupant

[80] A map showing the manner in which range control was exercised through ownership of alternate railroad sections may be found on the opposite page.

[81] United States *v.* Douglas, Willan-Sartoris, 3 *Wyoming Reports* 288, 1888.

of a given section, but no such occupation had occurred. The court laid the blame for such a condition to the Government's method of granting railroad land. In the opinion of the court

. . . . the grant was made in an unprovident fashion. The Government established the rectangular system of surveying, and so separated the odd- from the even-numbered sections as to leave between them only a geometrical line. It thus made the possibility of a clash of whatever conflicting interests might ensue between it and its grantees. . . . . Whose fault is it that the opposing parties are involved in this dilemma? If blame shall rest anywhere, it must lay at the door of the Government, which, being primarily the owner of the land, platted it and then granted it to the railroad company.[82]

Not until 1895, when a similar case from Colorado appeared in the United States Circuit Court of Appeals for the Eighth District, was the validity of the law of 1885 upheld and the practice of alternate section fencing declared an illegal enclosure.[83]

In the preceding pages, an effort has been made to present the problem of the control of the range as it appeared to the cattleman. Unlike the herdsman of the earlier frontiers, who moved on as the farmer took up the public domain, the cattleman of this last frontier proposed to succeed himself. First of all, he had no new frontier to move out upon; and second, he was not being threatened by any compact agricultural advance. As we shall see, the farmer's frontier was fraying out into isolated irrigation settlements, hopeless last stands on semi-arid desert claims, and small stock ranches whose owners warred with their larger and more powerful neighbors. What had been a temporary utilization of the unused public domain, the cattlemen of the eighties deter-

---

[82] *Ibid.*, 297.
[83] Canfield et al. *v.* United States; 66 *Federal* 101, 1895.

mined to make a permanent one. Such permanency could only be achieved through ownership of a part of the range, for without such private control, the industry would destroy itself by overcrowding.

It was in this process of shifting from the public domain as a physical basis of the industry, to the privately owned pasture and hay lands, that the stress came. It was obvious that the stockman could not acquire within the law, land enough to carry on the business as it had developed on the open range. It was equally obvious that the country as a whole would not consent to a change in the land laws that would suit the demands of the range country. The alienation of large, compact blocks of government land was completely at variance with the traditional land policy of the Republic. If this were permitted, the people of the United States stood a good chance of seeing the last of their estate slip into the hands of a few powerful companies, many of them controlled by foreign interests. Any move to enlarge the amount of land that an individual could obtain in these semi-arid regions was immediately met with the cry of land monopoly.

Finally, in spite of the investigation of land commissions and government agencies and of the proposals of the Land Office, there was not yet a sufficient body of knowledge on which to base any adjustment of our land laws to the far-western environment. No one could say with certainty what the future of those semi-arid regions was going to be. Only through the hard processes of trial and error were those methods of stock growing and agriculture going to develop which would result in the best utilization of the High Plains. The catastrophe that struck the cattle-growing industry in 1886-1887 demonstrated that the open range method of utilization could no longer be employed. The form of stock growing or agriculture, or a combination of both, that would best take its place was and still is a matter for speculation.

VII

DISASTER AND TRANSITION

IN spite of the efforts of some of the cattle companies to establish their business upon a sound economic basis through private ownership of the grazing grounds, the industry as a whole still rested in 1885 upon the open range; a basis that was becoming more and more insecure with every passing year. From the very first, the cattleman had had to reckon with winter losses, for, under the loose methods of handling stock prevailing on the open range, wintering cattle was "nothing less than slow starvation; a test of stored flesh and vitality against the hard storms until grass comes again." [1] In such a battle with the elements, the loss of 5 or even 10 per cent of the herd was not regarded as extreme. If, however, the stored vitality was reduced by poor summer feeding, due to drouth or crowded ranges, then no man dared to contemplate the disaster that a severe winter might bring.

The winter of 1880-1881 had been a severe one for the northern range, and in some sections, particularly in central Montana, the loss had been heavy. The succeeding winters were mild ones and the losses of 1880-1881 were soon forgotten in the fever of speculation that ensued. Cattle raising became more and more of a gamble, as one pioneer cattleman observed, "with the trump cards in the hands of the elements." [2] Higher and higher rose the stakes, and greater and greater became the risks. Herd was crowded in on herd till every square mile of pasturage was utilized. Hundreds of thousands of young eastern stock, purchased at high

[1] *Sen. Ex. Doc.* No. 16, 48 Cong., Sess. 2, p. 21.
[2] Stuart, II, 227.

figures, were thrown into this game of matching cattle against the weather. These eastern "pilgrims" exhibited a dangerous tendency to stand about the haystack in the winter, waiting to be fed, instead of rustling as did the native stock.[3] Each spring, there was a general counting of losses and a feeling of relief that fortune still favored the players of this desperate game. "More or less concern is always felt for the Northwest at this season of the year," commented a stock journal in March, 1885, "but from all accounts, there is nothing at present to cause uneasiness." [4]

This feeling of uneasiness, of approaching disaster, was strengthened by news of enormous losses on the southern ranges in the winter of 1885-1886. In the fall of 1885, the crowded ranges of western Kansas, Colorado, and the Panhandle were burdened still further by a flood of cattle arriving late in the season from the Indian Territory. The proclamation of President Cleveland, August 23, 1885, had ordered the cattlemen to remove their herds from the Cheyenne-Arapahoe reservation, where they had operated under leases, made a few years previous. Over 200,000 head were, by this edict, forced upon the overcrowded ranges, just when that area was about to experience one of the severest winters in its history.[5] Starvation and

---

[3] Lincoln Lang, *Ranching with Roosevelt*, 144.

[4] Quoted from the *Northwestern Live Stock Journal* in the *Bozeman Chronicle*, March 11, 1885. Bill Nye's humorous comment was not quite so optimistic. "Let me warn the amateur cowman," he wrote, "that in the great grazing regions it takes a great many acres of thin grass to maintain the adult steer in affluence for twelve months, and the great pastures at the bases of the mountains are pretty well tested. Moreover, I believe that these great conventions of cattlemen, where free grass and easily acquired fortunes are naturally advertised, will tend to overstock the ranges at last, and founder the goose that lays the golden egg. This, of course, is none of my business, but if I didn't now and then refer to matters that don't concern me, I would be regarded as reticent." *Laramie Boomerang* quoted in the *Bozeman Chronicle*, February 24, 1886.

[5] E. E. Dale, *History of the Ranch Cattle Industry in Oklahoma*, 315 *et seq.*

the blizzard did their work, and in the spring of 1886 the cattlemen in these regions found the carcasses of 85 per cent or more of their herds in the ravines or piled up along the drift fences.

The fall of 1886 found the cattlemen of Wyoming in a panicky condition. They were in the grip of a depression that would have caused a crash even though the winters had continued mild. Cattle prices in the Chicago market were steadily declining. The demand for stock cattle had disappeared and the rush of southwestern cattlemen to get out of the business had depressed the market. Cattle were selling for a lower figure than ever before in the history of the range, ten to fifteen dollars a head cheaper than in the preceding fall, with an overabundance of poorer grades, the result of the heavy turnoff of the weaker stock.[6] "Beef is low, very low, and prices are tending downward, while the market continues to grow weaker every day," warned the editor of the *Rocky Mountain Husbandman*. "But for all that, it would be better to sell at a low figure, than to endanger the whole herd by having the range overstocked." [7]

There was a general casting about for some means of protection. Many of the Montana cattlemen drove north across the Canadian border to the ranges of Alberta, where it was possible to lease large tracts of grazing land from the Provincial Government.[8] In August, 1886, 40,000 head from Dawson and Custer Counties alone were on their way.[9] By September, it was estimated that over a quarter of a million cattle from Montana would winter on the Alberta ranges.[10] Others looked to the unstocked Indian reservations for re-

[6] *St. Paul Pioneer Press,* quoted in the *Bozeman Chronicle,* November 24, 1886.
[7] *Rocky Mountain Husbandman,* August 26, 1886.
[8] Donaldson, *The Public Domain,* 485.
[9] The *Stockgrower's Journal* (Miles City), August 12, 1886.
[10] *Rocky Mountain Husbandman,* September 9, 1886.

lief. Would the central Government stand by and see cattle perish by the thousands, while the vast, ungrazed pastures of the reservations lay idle? Stockmen, on their way to Alberta drove across the northern reservation, league upon league of natural pastures, as empty as when the last buffalo left them. Deals were made with the Crow Agency, whereby several herds were permitted to winter on the reservation at the rate of fifty cents a head. This turned out to be an unprofitable venture, for the cattle were put on the reservation late in the season, and near the river, where the grazing was only fair. To the winter losses experienced by these herds was added the loss from killing by the Indians.[11]

Some of the cattlemen attempted to forestall disaster by boarding out a part of the herd. In Montana, the weaker cattle, cows and young stock, were collected in small herds of one to five hundred and driven southwestward into the agricultural section of the Territory where they were let out on shares to the small ranchers who had feed to spare. This return of the Montana stock growers to the older sections is significant, for it presaged the change in methods which animal husbandry in the Northwest was to undergo.

In Wyoming similar precautions were being taken. A Cheyenne paper noted the large number of cattle shipped in the fall of 1886 from the Wyoming ranges to the farmers in Iowa and Nebraska where they sold at better prices than in Chicago.[12]

All these precautions were eleventh-hour measures and did not to any appreciable degree break the force of the disaster. The summer of 1886 had been hot and dry, the grass had not flourished. Feed was short and the animals approached the winter in a poor condition. Even had there

[11] *Rocky Mountain Husbandman*, October 7, December 23, 1886, and February 17, 1887.
[12] *Cheyenne Daily Sun*, December 8, 1886.

been a mild winter, great loss would have been experienced in some sections, where there was not sufficient feed to bring them all through.[13]

In the latter part of November, there was a heavy fall of snow, so heavy that in many places the cattle could not get down to the grass. Gloomy reports began to come in from all sections. Those who had put up hay fed all they could, the rest whose cattle were all out on the ranges prayed for a chinook. It came, early in January, booming up from the southwest, melting the snow and blowing the exposed ridges bare. Men took heart, they might get through without disaster. But the odds were against them, for from the twenty-eighth of January to the thirtieth, the Northwest was swept by a blizzard such as the ranges had never before experienced. Down from the north, came a terrific wind before which the cattle drifted aimlessly or sought shelter in the coulees. A merciless cold locked up every bit of the poor grazing that remained. Men were forced to keep to the ranch houses for weeks as the bitter cold and the high winds scourged the range. They dared not think of the tragedy that was being enacted outside. Unacclimated "dogies" and young stock from Iowa and Wisconsin huddled in the quaking aspens and cottonwoods to die. Dry cows and steers, whose resistance was greater, lingered on. One morning the inhabitants on the outskirts of Great Falls looked out through the swirl of snow to see the gaunt, reeling figures of the leaders of a herd of five thousand that had drifted down to the frozen Missouri from the north. Inhabitants of ranch houses tried not to hear the noises that came from beyond the corrals. The longing for another chinook that never arrived became the yearning for a miracle. Old-timers, who were hardened to range losses, were in a state of absolute panic. From the Canadian range came the news that the

[13] *Rocky Mountain Husbandman*, October 14, 1886.

cattle there were faring as badly as those further south. The disaster was complete.[14]

Spring came at last, and the cattlemen rode out to face the reckoning. The sight of the ranges in the spring of 1887 was never forgotten. Dead were piled in the coulees. Poor emaciated remnants of great herds wandered about with frozen ears, tails, feet, and legs, so weak that they were scarcely able to move. Men revolted against the whole range system. "A business," wrote Granville Stuart, years afterward, "that had been fascinating to me before, suddenly became distasteful. I never wanted to own again an animal that I could not feed and shelter." [15]

The *Cheyenne Sun*, which had been a strong stockman's paper, in commenting on the condition in Wyoming in 1887, declared that "a man who turns out a lot of cattle on a barren plain without making provision for feeding them will not only suffer a financial loss but also the loss of the respect of the community in which he lives." [16]

It was felt that the catastrophe marked the end of an era in the history of the Northwest. "The fact that we have now to face," one observer reflected, "is that the range of the past is gone; that of the present is of little worth and cannot be relied on in the future. Range husbandry is over,

---

[14] The Wyoming and Montana papers were full of reports of the disaster. The above description has been written chiefly from them.

[15] Stuart, II, 237.

[16] *Cheyenne Daily Sun*, December 8, 1887. The cattlemen had for some years been subjected to eastern criticism, particularly in humanitarian circles, for cruelty in allowing cattle to go unsheltered and uncared for during the winter on the ranges. That they were sensitive to these charges is evidenced by the statement of Secretary Sturgis before the National Cattle Growers' Convention in Chicago in 1886. "The serious and oft-repeated charge of inhumanity is laid at the door of the ranchmen," he said, and added that ". . . . no part of the business is so constantly discussed. The reason for its long continuance is not indifference but lies in the immense inherent difficulties of the situation." *Cheyenne Daily Sun*, November 25, 1886.

is ruined, destroyed, it may have been by the insatiable greed of its followers." [17]

Those most responsible for the boom were the largest loosers. Large companies, which had bought great herds at high prices on borrowed money at high rates of interest and had thrown these herds upon the ranges as so many pawns in the game that was being played in the brokerage offices in Chicago and New York, disappeared overnight.[18] "Cattle barons" and "bovine kings" faded out of the public interest. English and Scotch investors tried to forget that they had been taken in by another sure thing. In May, 1887, the Swan Land and Cattle Company, largest of all the northern companies, went into the hands of the receiver with Willis Vandevanter, later justice of the Supreme Court of the United States, as receiver for the Swan interests.

There was a general unloading of the ranges as creditors demanded a settlement. In Wyoming alone this process of liquidation resulted in the shipment of between sixty and seventy thousand head of stock during the summer and fall of 1887.[19] In some sections these heavy shipments of the survivors practically cleared the ranges. Prices were ruin-

[17] *Rocky Mountain Husbandman*, March 17, 1887.

[18] Stuart, II, 236. The locality found a certain satisfaction in the disappearance of some of these companies. The *Cheyenne Daily Sun* congratulated its readers on the departure of the Frewen brothers, managers and promoters of the Powder River Cattle Company: "Of all the English snobs of great pretentions, who flew so high and sunk so low, probably the Frewens are the chiefs. Their careers in Wyoming as cattle kings will long be remembered. They . . . . made cowboys of freshly imported lads from England, maintained a princely establishment on the frontier, ... established relay stations so as to make lightning journeys through the territory; had flowers shipped to the ranch, and conducted business . . . . on a system . . . . that was a constant surprise even to the most reckless and extravagant Americans. It is this method . . . . that has brought an important and legitimate business into discredit in the East." *Cheyenne Daily Sun*, November 3, 1887.

[19] *Cheyenne Daily Sun*, November 1, 1887.

ously low in Chicago, but the turnoff continued until the deflation of the stock industry of the northern ranges was complete.

There was some reason for believing that the disaster that the range-cattle industry had suffered was only a temporary setback, and that a few good years would bring it back to its old position of power. In the first place, the range was there, and although it had been overgrazed in certain sections, the relief that the disaster and deflation had brought would soon restore it to its former value. Furthermore, the reduction in numbers gave to those who still remained in the business an opportunity that had not been theirs since the early days of the range. It would now be possible to distinguish between summer and winter feeding grounds, reserving the latter for the winter season, a practice that had not been followed during the boom period, when winter range was often fed off before the summer was over.

In the spring following the disastrous winter of 1887, Nature was never more favorable to the northern ranges. The heavy snow, which had brought death to so many thousands of range stock, furnished plenty of moisture to grow an excellent crop of grass for the survivors. When winter came around again, the cattle were in good flesh to meet it. A favorably placed chinook, which arrived in February, 1888, proved opportune and the cattle came through in good shape.[20]

Another circumstance seemed to point to the early recovery of the northern ranges. In 1888, the enormous Indian reservation in northern Montana was reduced in size by cutting it up into three small reservations. A large area of untouched grazing ground was thus opened up, where, it

[20] *Rocky Mountain Husbandman*, December 8, 1887; February 9, 1888.

was believed, the cattle industry of Montana might continue on the old range basis.[21]

In spite of these favoring conditions, the range-cattle industry was on the decline, although, in some quarters, it lingered on for more than a decade longer. In the first place, the old confidence in the range was gone. Never again would cattlemen dare to take the chances that had been regarded as part of the business in the earlier day. Those who still remained in the business found the margin of profit so small that a winter loss that had been but an average one in the old days would now prove ruinous. The range no longer appeared a safe basis for the industry.

Outside capital, so plentiful during the boom period, was now no longer available. In the disaster, the larger companies had suffered the most; many had gone under and of the few that had survived, none were free from enormous liabilities incurred during the era of speculation. Only by the most careful management could the survivors hope to continue in business, and as for any return on the investment in either cattle or land, it was likely to be small for some time to come. Cattlemen and stock journals no longer talked of easy profits and rapidly expanding herds, but rather of the necessity of reducing the size of operations, buying land, and putting up hay. The day of the beef bonanza was gone; investment in the western range-cattle business was no longer attractive.

Finally, the general depression of the late eighties and early nineties made the recovery of the cattle industry in the Northwest a very slow process. Prices continued low, for the market could not take up the turnoff during this period of deflation. From the northern ranges, shipments continued

[21] The reduction was agreed to by the Indians, December, 1887, and ratified in a law approved May 1, 1888. *Indian Laws and Treaties,* I, 261-266.

to be heavy as the survivors pared down their herds to the safety point or sold out altogether and turned to raising sheep. Texas growers, finding it more and more difficult to sell to the northern cattleman, were forced to add to the glut of cattle on the eastern market. To these, was added the stock of the farmers of western Kansas and Nebraska. In these regions, a series of dry years had caused general failure, a recession of the over-extended farmers' frontier, and a disposal of the farm stock for whatever it would bring. In 1889 more than three million head were received in Chicago from these various quarters, of which over 25 per cent were cows and 4 per cent calves.[22]

In the face of such conditions, it was futile to hope for a return to the old range days. Ten years after the disaster, the condition in Wyoming shows how completely the range cattle industry was passing out of the picture. By 1895, there had been a shrinkage in the assessed value of cattle of over a million dollars a year. Numbers had declined from nine million head in 1886 to less than three million head in 1895. So that the number of cattle in Wyoming in 1895 was about what it had been in 1880 when the northern ranges were just beginning to be opened up.[23]

But the open range, which in 1880 had been the basis upon which the profits of the boom period had rested, was being only partially used in 1895. It was still there, over 80 per cent of the total area of the State being in public domain in 1890.[24] Some of it, it is true, was being utilized by the sheepmen, whose bands had increased in valuation during the ten-year period by about two million dollars. As

[22] Vest Report, p. 8.

[23] *First Annual Report of the Wyoming State Engineer*, 1895 (Cheyenne, 1895).

[24] *Annual Report of the Commissioner of the General Land Office*, 1890, p. 121.

for the remaining grazing resources of the open range, the stock grower, who was developing a different system, looked upon their use as only incidental.

The development of new methods of plant and animal husbandry in the northern arid regions coincided with this decline in the use of the open range. In this period of transition, it is possible to distinguish several economic groups, although as we shall see, they tend to merge one into the other at various stages in the process of adjustment. There were first, the few remaining large cattle companies, many of them owning large blocks of land; second, a number of small cattle growers whose stock, usually in small bands and close herded, utilized part of the natural grazing resources; third, a group of irrigators, occupying favorable spots; fourth, the "grangers" or frontier farmers on the fringe of farming settlement, who had arrived in eastern Montana and Wyoming by the close of the range period; fifth, the older agricultural communities of the more fertile valleys of western Montana, where the moisture was sufficient to carry on farming operations with a minimum of irrigation; and sixth, the sheep owners, large and small. These various groups were the successors of the old range cattlemen of the eighties.

The winter of 1886 and 1887 had demonstrated to large and small cattleman alike that to depend solely on the grass crop of the open ranges had been a fatal mistake. The future safety of the industry lay in the production of hay sufficient to carry the herds through the winter at least. We have noted the tendency in that direction before the catastrophe. Some of the larger companies had purchased wild hay lands, wherever available, and had made some effort to provide feed for the weaker stock during the most severe weather. In central Montana, particularly in the Smith River region, where the newer range industry came in touch with the older farming communities, hay lands were being fenced and cropped and cattle cared for on a close-herding basis. Had there been no disaster, the drift toward land ownership in order to control a part of the grazing would have eventually forced the substitution of forage crops for the native grasses. The attendant costs of land ownership, taxes, interest, and fencing, would have brought about a condition where the production of hay wherever possible would be a more economical utilization of the land than using it as natural pasture for part of the time and grazing the open range for the remainder. When this point was reached, pasturing on the open range, with its attendant dangers and costs of herding, would be discarded; first, the winter and then the summer pasturing.

The disaster merely hastened the process. The larger range companies, which had some land, led off. The Douglas correspondent of the *Northwestern Live Stock Journal* in March, 1887, noted the arrival in town of the manager of one of the large cattle companies on the Powder River with a shipment of twelve thousand pounds of grass seed — red top, timothy, alfalfa, clover, and blue grass.[25] The change suggested by the appearance of such items in a range journal was, of course, gradual, but none the less, fundamental. This reorganization on a new basis has been well described by a manager of one of the old companies:

It was evident that the old system of the open range was passing and this was intensified by a series of dry years [1890 was very dry]. Both grass and hay were short, but with good winters we managed to pull through. We therefore commenced to get quit of all frills and

[25] *Northwestern Live Stock Journal*, March 18, 1887. Alfalfa, now the chief forage crop of the Northwest, was introduced into western Montana, Madison County, around 1880. The *Rocky Mountain Husbandman* in 1887 urged its readers to experiment with it in place of the standard forage plant, timothy. Local merchants were purchasing seed from Colorado, Utah, and California. *Rocky Mountain Husbandman*, March 17, 1887.

come down to earth, as the cowboys express it. We rented out our ranches . . . . paying so much per ton for hay for every one put in the stack and then we paid the renters so much a month in winter time for feeding out the hay, cleaning out ditches, building fences, etc. They grubbed themselves and we paid so much a meal for any of our help that had to stay at their places. We produced immensely more hay, eliminated waste, and the system worked like a charm. As we raised hay and improved our pastures, we began weaning calves, feeding cows and heifers, while away on the Plains in Bate's Hole, in rough country north of the U. P. toward the Platte River, our steers took their chances.

In 1888, this company put up 1,200 tons of hay, in 1892, 3,600 and by 1904, 14,600 tons.

While it does not take long to write the account of these changes, it was a great task to reorganize the machinery of ranch work, cut down the cattle outfits, till in spring we had but one wagon [chuck wagon] running, with the help of a second in the fall, and look after the feeding of twelve to fifteen thousand head of cattle in pasture. As the work contracted, we had to face blackleg and destruction by wolves among our calves, and more or less mange appeared among our cattle. Most of our neighbors moved out and left us. Sheep trespassers on our plains lands were hard to handle, and we had a continual fight on our hands. Eventually, the courts granted injunctions and this range graft was stopped.[26]

The acreage devoted to the cultivation of hay in Wyoming and Montana increased more than tenfold during the period from 1880-1900, as the census figures for the three periods show:

Date	Montana Acres Mown	Wyoming Acres Mown
1880	56,801	24,328
1890	300,033	173,010
1900	712,048	377,148

These figures not only mark the decline of the range-cattle

[26] Clay, 222-224.

industry, but also the passing of that distinctive frontier figure, the cowboy. Mowing machines, hay rakes, and ditching tools became as important a part of ranch equipment as the chuck wagon, the lariat, and the branding iron.

Cowboys don't have as soft a time as they did [one member of the fraternity lamented]. I remember when we sat around the fire the winter through and didn't do a lick of work for five or six months of the year, except to chop a little wood to build a fire to keep warm by. Now we go on the general roundup, then the calf roundup, then comes haying — something that the old-time cowboy never dreamed of — then the beef roundup and the fall calf roundup and gathering bulls and weak cows, and after all this, a winter of feeding hay. I tell you times have changed. You didn't hear the sound of a mowing machine in this country ten years ago. We didn't have any hay and the man who thinks he is going to strike a soft job now in a cow camp is woefully left.[27]

Some of the larger operators turned to sheep raising. In Wyoming the rise of the sheep population between 1890 and 1900 was very marked.[28] Like the cattle growers, the sheepmen began to combine the summer pasturage of the open range with the winter feeding of hay, raised on privately owned or leased land. One Wyoming sheep company reported in 1889 its holdings as follows:

	Acres
Land in fee simple	96,000
Leased University and school land in Wyoming and Colorado	23,000
Range rights	150,000
Government land	15,000
Total	284,000

The portion of this ranch lying south of the Union Pacific

[27] Quoted from the *Barber County* (Kansas) *Index* in the *Breeder's Gazette*, VI, 608, October 23, 1884.

[28] See page 230.

was described as being twenty-five miles long and seven miles wide, all fenced, partially irrigated by thirty miles of main ditch and sixty-five miles of laterals. Eighteen hundred tons of hay were being cut yearly to feed the flocks, which numbered about seventy thousand head. The company main-

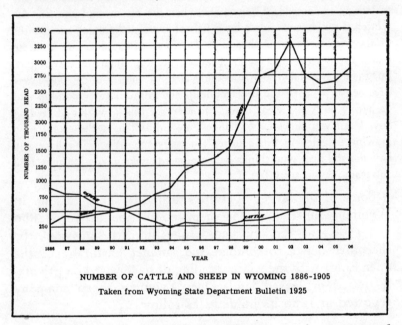

NUMBER OF CATTLE AND SHEEP IN WYOMING 1886-1905
Taken from Wyoming State Department Bulletin 1925

tained thirty-eight ranch houses and sheep stations scattered over this area, connected one with the other by telephone.[29]

The development of irrigation was closely associated with the shift of the stock-growing industry from the open range to the ranch basis. During the early period of settlement, the raising of grain crops, wheat, oats, etc., in Montana and Wyoming had been limited to a few fertile areas found along the rich bottom lands where the fertility of the soil and the greater moisture close to the mountains made

[29] Advertisement of the Warren Live Stock Company in the "Statehood Edition" of the *Cheyenne Daily Sun*, March 28, 1889.

production without irrigation possible. This type of farming was carried on in the Beaverhead, Madison, Deer Lodge, and Gallatin valleys in Montana and along the upper reaches of the Wind River Valley in Wyoming. The next step in the extension of the farming area came in those spots where the contour of the land and the availability of water made irrigation comparatively easy. Miners, turned ranchmen, who had learned how to get water on their mining claims by running ditches and flumes along the sides of the gulches, began to put this knowledge to work in the narrow valley lands upon which some of them settled after the mining boom was over. Occasionally, a project that required more than the effort of a single rancher was attempted, and a ditch company was formed. The incorporation list of the Territory of Wyoming shows how limited this development was in the seventies as compared with the succeeding decade:[30]

Date	No. Companies Incorporated	Date	No. Companies Incorporated
1870	1	1879	3
1871	1	1881	3
1874	2	1882	11
1875	1	1883	19
1876	1	1885	36

By 1890, four types of holdings were distinguishable on the northern ranges. There was first the unirrigated farm, which, until the development of dry farming, was limited to those few favorable localities in the mountain valleys; second, the unirrigated stock ranch, which still depended upon a combination of natural hay-lands and open range; third, the irrigated farm, raising some crop other than forage; and fourth, the irrigated stock ranch where irriga-

---

[30] *Annual Report of the Governor of Wyoming to the Secretary of the Interior*, 1885 (Washington), 76.

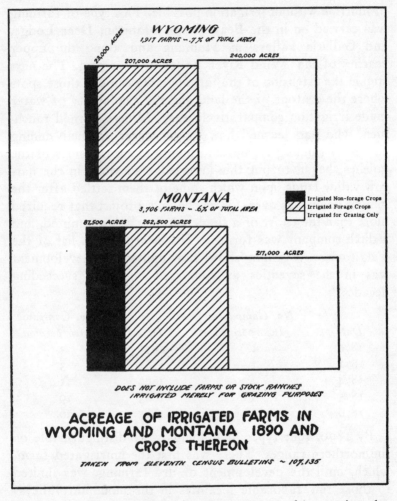

WYOMING
1,917 FARMS ~ .7% OF TOTAL AREA

23,000 ACRES

207,000 ACRES

240,000 ACRES

MONTANA
3,706 FARMS ~ .6% OF TOTAL AREA

Irrigated Non-forage Crops
Irrigated Forage Crops
Irrigated for Grazing Only

87,500 ACRES

262,500 ACRES

217,000 ACRES

DOES NOT INCLUDE FARMS OR STOCK RANCHES
IRRIGATED MERELY FOR GRAZING PURPOSES

ACREAGE OF IRRIGATED FARMS IN
WYOMING AND MONTANA 1890 AND
CROPS THEREON

TAKEN FROM ELEVENTH CENSUS BULLETINS ~ 107,135

tion was carried on solely for the purpose of increasing the forage crop on the natural pasture lands.

The first and second group have been previously described and represent the older type of agriculture and stock raising with which the reader is now familiar. On the ranches of the third and fourth groups, irrigation was being employed either in the production of farm crops, cereals and the like, and forage, or in the improvement of the natural pastures. In the third group, crop production under irrigation was being combined with stock raising; in the fourth, stock raising alone was carried on, with irrigation employed merely to increase the crop of grass found on the natural pastures. In both groups, however, stock raising and irrigation were combined, and this alliance is one of the most important features in the process of adjustment following the range-cattle days.

Even in the third group, where cereals were being raised under irrigation, the area devoted to these crops was very small compared with the acreage irrigated for forage crops or for the improvement of the natural pastures. The accompanying diagram shows the preponderant importance of stock raising in this group.

As for the fourth group, stock growers irrigating pasture lands, it is impossible from the available census figures to tell how many there were or the area of the land they occupied, as the census does not give the figures for any group of irrigators except those who were irrigating not solely for grazing purposes. They were not irrigators as the term is commonly understood but represented a transitional group, who were moving over from the old range-cattle group to the stock grower with forage crops raised by irrigation.

The irrigation systems developed by this latter group were of the crudest and most temporary character, for they were only designed to improve the natural grass crop in favorably located spots. In a report in 1890 on irrigation in the western states prepared for the Eleventh Census, the investigator described this irrigating practice as follows:

Many canals, especially along the larger rivers, receive water only in time of flood. . . . . These so-called high-water canals are very

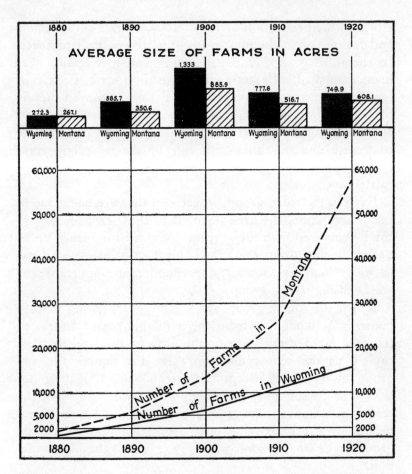

cheaply built, the headworks, if any, being of a temporary character. Water is by this means turned out upon the grazing lands; only one thorough watering during the year being secured.[81]

As time went on, the rising costs of land ownership forced these stock growers to a more complete utilization of their

[81] F. H. Newell, "Report of Agriculture by Irrigation in the Western Part of the United States at the Eleventh Census, 1890." *House Mis. Doc.* No. 340, Part 20, 52 Cong., Sess. 1, p. 249.

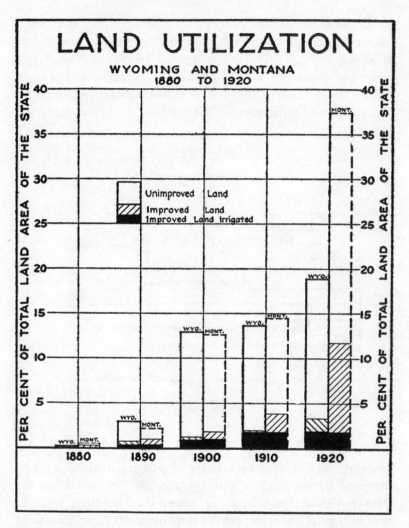

land by planting forage crops instead of merely seeking to improve the natural grasses. Thus, this group tended to disappear into the third group and irrigation systems to assume a more permanent character.

This alliance of irrigation with stock raising can be traced in the increase of the average size of farms in the range country. Between 1880 and 1890, the average size of farms in Wyoming and Montana increased from 272.3 to 1,333 acres and from 267.1 to 885.9 acres, respectively. See page 234. This increase in the size of farms was accompanied by a corresponding increase in the number of farms, as stock

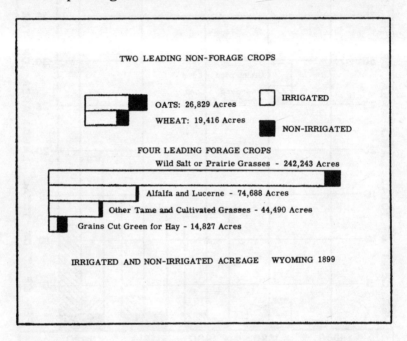

growing moved over to a basis of private ownership. The decrease between 1890 and 1900 in the improved but un-irrigated land shows how the newer land holdings were of that marginal character where improvement must consist of the development of irrigation. Forage crops continued to be predominant, as the stock-growing industry shifted to an irrigated forage crop basis. So that by 1900 the land held was chiefly irrigated land or unimproved pasture, which, looked at from a stock-growing point of view, represented

haystacks for winter and enclosed pasture for summer grazing. See pages 235 and 236. Following 1900, the increase in
the improved, non-irrigated land, first in Montana and then
in Wyoming, marks the arrival in force of the dry-land
farmer. This new frontier figure, in his struggle to conquer
a part of the range that the cattleman had abandoned in
defeat, has brought the story of the utilization of the semi-
arid West down to the present day.

This retreat of the stock-growing industry behind the
defenses of privately owned pastures and irrigated forage
crops was not immediate. The attempt to utilize the open
range continued for some time to come. In some regions,
stock growers who were raising feed for winter continued
to use the range for summer pasturage, in others, the old
open-range methods lingered and cattle, particularly steers,
were wintered on the gradually contracting ranges. In these
areas, the older cattlemen and companies put up a rear-
guard action against sheepmen, small cattle owners, and
grangers. On the northern ranges, and particularly in
Wyoming, this clash between the surviving cattle companies
and the small ranchmen and grangers constitutes the final
chapter in the story of the range-cattle industry.

Even before the catastrophe in 1886-1887, the northern
cattle country was conscious of the arrival in western Nebraska and eastern Wyoming of new elements of the population different from, and to a large degree antagonistic to,
the cattleman. There was not, however, any compact advance of a westward-moving farming frontier such as swept
the Kansas cattleman out of the picture. Rather, there appeared first a number of small ranchers whose stock mingled
on the open range with those of the larger owners. The advertisement that the cattle country had received during the
boom period brought out these small operators as well as the
larger companies. To these were added cowboys who had

left the employ of others and were building up small herds of their own.[32]

Antagonism between the small rancher and the larger cattle outfits soon developed. Although the associations were sensitive to the charge that they were inimical to the small cattle grower and sought to demonstrate the fundamental democracy of their organizations, the fact remained that the owner of a small but growing herd was, from the very nature of the case, looked upon with suspicion, often founded upon the experience of some of the older members in establishing a herd in an earlier and less regulated day.[33] If it was difficult for the associations to distinguish the small cattleman from the cattle thief, it was becoming equally difficult for newcomers to escape the conclusion that an association like that in Wyoming, which represented the whole range-cattle business in that section, might easily use its power to dominate the cattle range and to exclude all but a few wealthy companies.

The shift of some of these powers to the territorial government did not tend to allay suspicion. In 1888 the Wyoming legislature created a Board of Live Stock Commissioners with power to regulate roundups, sell mavericks, and direct the inspection, powers which the Association had

[32] Clay states that there were 3,500 brands in Wyoming in 1891 (p. 264). The *Cheyenne Leader*, Dec. 6, 1891, put the figure at 5,000. In Montana, the Live Stock Commission listed about 6,000 in 1889; by 1892 this had risen to 11,000; 1900 to 16,000; and 1910 to 20,000. These figures have been taken from the annual reports of the Secretary of the Board of Live Stock Commissioners of Montana for the respective years.

[33] The Secretary of the Wyoming Association in 1884 in a newspaper interview declared that the small stockman was as welcome in the Association and had as much power in voting as Swan or Carey. Eighty members out of more than four hundred owned less than a thousand head in 1884; one, who had been a member for years, having only twenty-five head. (*Cheyenne Daily Sun*, January 30, 1884.) At the spring meeting in 1892, of the forty-three names on the roll call, all but eight were either owners or managers of large cattle companies. *Cheyenne Daily Leader*, April 5, 1892.

exercised since 1884.[34] The catastrophe of 1887 had so weakened the Association, both in prestige and financial strength, that it was glad to turn the task over to the Territory. The men appointed by the governor to the newly created board were, without exception, leading members of the old Association, thus perpetuating the alliance between the cattlemen and the territorial government.

As for the Association itself, the decline in membership after 1887 necessitated a policy of retrenchment and contraction.[35] Salaries were cut, and detectives and inspectors dismissed.[36] In some cases, detectives were re-employed, ostensibly by the Association, but actually by a few of the large companies to watch out for their particular interests at strategic points.[37] It was hoped that the organization, now composed almost exclusively of the managers or owners of the surviving companies, might be kept alive by the payment of a small membership fee only. Assessments were done away with in 1890; for it was thought that the Live Stock Commission would be able to pay for inspection.[38]

The newly created Commission soon found itself in serious difficulty. The only money that it had to pay for carrying on the work it had taken over from the Association was the receipts from the sale of the mavericks. This source of

[34] *Laws of Wyoming Territory*, 1888, pp. 46-54.

[35] In 1888 the total membership list was 203, many of these being old members who were no longer in Wyoming but still kept their membership.

[36] Executive Minute Book of the Wyoming Stock Growers' Association, p. 70, Aug. 7, 1887; p. 76, Dec. 10, 1888; p. 78, Apr. 14, 1889.

[37] Smith, the Association inspector at Chadron, Nebraska, was re-employed by four of the large cattle companie* and directed by the Secretary of the Association to take his orders from Harry Oelrichs of the Anglo-American Cattle Company. Sturgis to Smith, Jan. 4, 1887. Correspondence of Wyoming Stock Growers' Association, Letter Book 7, p. 886.

[38] Executive Minute Book, p. 90, April 29, 1890. They were renewed in a year or so, when it was found that the Live Stock Commission was incapable of performing efficient inspection service.

revenue was in the way of disappearing altogether, for with the removal of the strong hand of the Association, cattle stealing and mavericking flourished as never before. Although the Commission was completely dominated by the Association, and desirous of affording protection to Wyoming's oldest industry, it was powerless. Indeed, the situation was so bad by the spring of 1891, that no man's cattle were safe. As for the big outfits, their herds were fair game for all. In commenting on the anarchy that ruled the ranges, a local Cheyenne paper, not overly sympathetic to the large cattle interests, declared that cattle stealing had become almost respectable.

Men who before this year have borne and deserved good characters are now openly engaged in preying upon the public ranges. . . . . All their neighbors and acquaintances are perfectly aware of the fact and the practice is oftentimes not merely winked at, but applauded. . . . . Efforts have been made by some of the larger cattle companies to bring the offenders to justice. In some cases the grand juries have refused to indict; in others petit juries have brought in verdicts of not guilty in the face of evidence as conclusive and convincing as any ever submitted in a court of justice. . . . . Small ranchmen have been terrorized into submission. They take good care to avoid seeing what is going on under their very noses and the reasons for doing so must be very obvious to everyone who knows anything of frontier life. . . . . There are only two horns to the dilemma, either the thieves or the cattlemen must go.[39]

Not since the disaster in 1886-1887 had the ranges been in better condition, the paper declared, but prophesied that unless some action were taken only the thieves would be left to possess it.[40]

Behind this screen of small ranchmen, who appeared to be laying the foundation of their future prosperity at the ex-

[39] *Cheyenne Daily Leader*, July 25, 1891.
[40] *Ibid.*, July 26, 1891.

pense of the big outfits, the small farmers or grangers were filtering in. In 1886, the Fremont, Elkhorn, and Missouri River Railroad, a subsidiary of the Chicago and North Western, had crossed the Wyoming line and by the following year had reached the North Platte at Douglas, a few miles below Fort Fetterman.[41] Along the line of this road, the farmer began an invasion of the northern ranges. Railroad advertisements, local papers in the small towns, and eastern farm journals created a mirage of farm homes out on the arid bench lands. A Democratic administration, pledged to find homes for poor men, declared that the farmer's frontier had not been stayed in its westward course by natural obstacles but by those "corporate cormorants," the cattle companies, who had been allowed to batten on the public estate under the complacent Republicans. The Democratic Secretary of the Interior in 1886 denied that the land laws were failures in the arid West and declared that, "These plains and plateaus are permanently adapted for the homes and husbandry of poor men." [42] Land Commissioner Sparks, zealously vocal, rang the changes on "homes for poor men" and "the robbing cattle barons." The newly arrived immigrant farmer from northern Europe took the Government at its word and proceeded westward to make the cattle ranges safe for democracy.

In the mass, Secretary Lamar's "poor men" were a serious threat to the range cattleman, second only to the cattle rustler. As individuals, however, they were more often objects of amused contempt or of a certain good-natured charity. The inspector of the Wyoming Stock Growers' Association, located in Sioux County, Nebraska, described those who had settled in the neighborhood as "an ignorant and degraded gang of continental paupers whose only stock

---

[41] *Poor's Manual*, 1886, p. 842; *Official Railway Guide*, June, 1887, p. 321.
[42] *Annual Report of the Secretary of the Interior*, 1886, p. 40.

in trade consists in a large number of ragged kids." [43] But many a "Dutch pauper" and his family of "ragged kids" were carried through the winter on provisions and beef furnished by a disgusted but soft-hearted foreman of a nearby cattle outfit.

By 1887 the grangers had arrived in force in western Nebraska, where lay the range of some of the largest owners in the Wyoming Association. The cowboys found them in little groups, plowing up their quarter-sections and planting crops. Fences were too expensive and wood too scarce, so the granger turned a furrow around the margin of his homestead, and let it go at that. When the range cattle got on his land, he shot at them, thereby protecting his land and furnishing his family with beef. Cattle owners might insist that the fence laws of the state required the farmer to fence his crop and might attempt to punish such destruction, but the killing went on just the same. Writing in March, 1887, to the Secretary of the Wyoming Association, the manager of the Anglo-American Cattle Company, operating in eastern Wyoming and western Nebraska, declared,

The loss on Hat Creek through the farmers has been something frightful. They have not confined themselves to killing for their own use, but have killed wantonly. On one stretch of a mile and a half on Hat Creek, there are over thirty cattle dead, every one of which shows rifle or pistol wounds.[44]

At the county seats, the granger influence grew, and soon it was next to impossible to get convictions for violations of the stock laws. Writing from North Platte, where the granger had appeared a few years earlier, one of the old-time stockmen complained to his Association that "Meth-

---

[43] Letter, Smith to Sturgis, Mar. 29, 1887, Correspondence of Wyoming Stock Growers' Association, Letter File, January-March, 1887.

[44] Harry Oelrichs, president of the Anglo-American Cattle Company, to Sturgis, Mar. 3, 1887, Correspondence of Wyoming Stock Growers' Association, Letter File, January-March, 1887.

odists, Grangers, and Anti-Stock, etc., were too many for us — they crept in on the Grand Jury and prevented any bill being brought against the accused parties." [45]

By 1888, Wyoming ranchers were feeling the full force of this invasion. The ranch owner who lived near the line of the newly constructed Chicago North Western found himself keeping a sort of road house for Germans, Russians, and Swedes on their way to their new homesteads. "They pile in on me at the rate of fifteen a day . . . . and all seem confident that they can make a success at farming," wrote a correspondent to a Cheyenne paper.[46] Hat Creek, Running Water, Shawnee Creek, and Goshen Hole, all famed stock ranges north of the Platte and between Fort Fetterman and the Nebraska boundary, were being settled up.[47] Running Water, long a station on the road between Fort Laramie and the Black Hills, became the town of Lusk in 1887, the center of a considerable granger population. Further on up the Platte, Douglas, a granger town, was settled in May, 1886, and by June had several hundred people.[48]

To the pioneer cattleman, "farm homes for poor men" on the Wyoming ranges was a romantic absurdity. "It is not a poor man's country," declared the leading stock journal of the Territory.

[45] John Bratt to Sturgis, June 12, 1884, Correspondence of the Wyoming Stock Growers' Association, Letter File, May-August, 1884. A member of one Wyoming jury wanted to know before he made up his mind on a case involving cattle killing, whether the cow had been killed to eat or to sell! *Cheyenne Daily Leader*, September 6, 1891.

[46] *Cheyenne Daily Sun*, Mar. 31, 1888.

[47] *Cheyenne Daily Sun*, April 21, 1888, May 25, 1888.

[48] *Ibid.*, June 18, 1886. Looking back on his experiences, one of the early cattlemen writes as follows of the "granger invasion," "Then the Northwestern Railroad came poking in and brought the festive granger. Then trouble did begin. It was not like the Indians, for one couldn't shoot and the only way I could get even, was to go into the banking business." C. F. Coffey, *Letters of Old Friends and Members*, 28.

The water of every stream in the Territory, if all utilized, is not sufficient to irrigate the first and second bottoms contiguous thereto, leaving the highlands forever unfavorable to agriculture. One-fifth of the area is rich first and second bottom lands, but save a small area of the mountain valleys, all have to be irrigated, in order to produce a crop, even of hay. The cost of irrigating, when water is free in the creek nearby, usually amounts to from five to ten dollars an acre; uphill business for a poor man. The valleys are usually small and settlements must always remain scattered. There is no timber save in the mountains and building material is high. There is no work for a poor man to do except to run cattle a few months in the summer for the stockmen. The little towns all over the Territory are full of idle men every winter. . . . . The time has passed when a man could invest a few thousand dollars on the Plains and make a vast fortune in a few years.[49]

Valid as this appraisal was, and later events served to demonstrate its essential accuracy, it coincided all too well with the desires of the cattlemen to preserve the ranges for their own use. If the American farmer desired to attempt the conquest of the semi-arid West, then the "cattle kings" must make way. A Democratic "anti-stock" governor, one Moonlight, appointed in 1887 by Cleveland, reported with great satisfaction that the old cattle companies, who conducted their business "in a lordly and magnificent manner," were on the road to ruin and that the small ranch, where stock raising and farming were combined, was succeeding them.[50] Stockmen were accused of "giving a black eye to the industry [agriculture] whenever an opportunity offered and . . . . [discouraging] anyone from engaging in it, as nothing hurts them so much as to see a cabin and a fence on the range." [51]

[49] *Northwestern Live Stock Journal* (Cheyenne), Jan. 21, 1887.

[50] Annual reports of Governor of Wyoming to the Secretary of the Interior, 1887 and 1888. The difficult problem of combining stock raising and agriculture on the High Plains, which appeared so simple to a newly arrived governor, has still to be solved.

[51] *Great Falls Tribune*, May 21, 1885.

Populism, which had arisen out of middle-western agrarian discontent, began to appear on the Wyoming ranges in the form of "anti-stock" agitation. In Kansas, the farmers with a grievance compounded of no rain, low prices, high interest rates, and high freight charges had ceased to raise corn and had started a crusade against the "big corporations," and the "money power." In Wyoming, the grangers took to shooting range cattle, sitting on grand juries, which boasted of never indicting an accused cattle rustler, and talking about the greed of the cattle companies.

The repeal of the preemption and timber culture laws, and the modification of the Desert Land Act appeared to them to be the work of the tools of the big corporations. After allowing the "cattle kings" to get all they desired, the Government now permitted the *status quo* to be preserved, by reducing the settler to a mere 320 acres of desert land, which he could not possibly irrigate. "Had the 'cattle kings' at whom the acts in question were supposed to be aimed, drawn a bill especially in their own interests, they could not have done better," declared the editor of the *Cheyenne Leader* in 1891.[52] The same paper carried stories, told under oath, of Senator Warren's sheep herders driving out the small settlers in the neighborhood of his vast range, or forcing them to sell out at ridiculously low figures. No wonder the senator voted for repeal and modification.[53] An indiscreet remark by the Secretary of the Board of Live Stock Commissioners that "there are too many people here now — too many people and not enough cattle," gave the anti-stock forces the opportunity to accuse the Wyoming Republicans of raising the standard of "More steers and fewer men." [54]

[52] *Cheyenne Daily Leader*, April 9, 1891.
[53] *Ibid.*, Oct. 11, 1892.
[54] *Ibid.*, July 12, 1892.

Beset by an "anti-stock" brand of Wyoming populism and by the cattle rustler with whom the granger made common cause, the old elements that had ruled Wyoming for two decades rallied to defend their position. In the fall of 1891, they took the offensive. The old Association gathered its strength for one last effort to save the range-cattle industry. The first line of attack on the rustlers was through the Live Stock Commission, which was completely dominated by the Association. In October, they instructed the inspectors in their employ at the markets to hold up all cattle bearing a brand known to be that of a cattle rustler. A list of rustlers' brands prepared by the Commission was forwarded to the inspectors. They were told to disregard all bills of sale held by the owner of cattle thus branded, as the rustlers were known to use the granger as a "fence" by giving him a bill of sale to prove ownership. These suspected cattle were to be sold by the inspectors and the money turned over to the Live Stock Board.[55] The cowboy, suspected of rustling, the small rancher who was unfortunate enough to incur the enmity of a neighboring large cattle outfit, and the granger with suspected stock to sell, were all hit by this action. They must come before an *ex parte* board and prove that the suspected stock belonging to them had not been stolen and that they were not rustlers. Such action by the Board, so at variance with the fundamental principle of Anglo-Saxon justice, could not fail to inflame public sentiment and bring

[55] *Cheyenne Daily Leader*, Oct. 28, 1891. The total number of strays checked at the inspection points in 1891 amounted to 16,306. Of these, 5,238 were sold and the money remitted directly to the Live Stock Commission. The proceeds from the sale amounted to $127,243.36 and of this amount, $113,293.53 was sent out in checks to the owners. The remaining $13,949.83 was held by the Commission. This last sum represented the money received for the sale of cattle belonging to alleged rustlers. Minute Book of the Wyoming Board of Live Stock Commissioners, p. 6. This book is in the possession of the secretary of the Board in Cheyenne, Wyo.

the honest ranchman and granger into line with the cattle rustlers.

The territorial press, which in the old days knew what it was to challenge the power of the Association, began to line up on the popular side. Threats to deprive the papers of advertising revenue were of no avail and blank spaces, once filled with advertisements of cattle banks, Association attorneys, and such mercantile firms as were subservient to the cattle companies, merely served to make the paper more popular.[56]

"We regret . . . . the revival of the old spirit of intolerance in the Wyoming Stock Growers' Association," commented the editor of the *Leader*, one of the papers thus attacked. "An un-American spirit of dominance which would ride roughshod over the weaker elements and force them to immigrate or crawl, cowed and subdued, at the feet of a fierce and implacable oligarchy."[57]

In November, 1891, matters took a more serious turn. News came down from Johnson County, in the northern part of the State, that two of the residents of that section had been shot from ambush. The two murdered men had been suspected of rustling and consequently the assassinations were immediately connected with the one or two large cattle companies in that neighborhood. A United States deputy marshal, formerly employed as an inspector by the Association, was suspected. Crowds gathered in Buffalo, the county seat, and threats were circulated against this officer and the chief cattleman in the region. The former was arrested by the local sheriff, but was able to give an alibi sufficient to obtain a release.[58]

Johnson County was in open revolt by the spring of 1892.

[56] *Cheyenne Daily Leader*, Mar. 22, 1892.
[57] *Ibid.*, Mar. 23, 1892.
[58] *Ibid.*, Dec. 6, Dec. 20, Dec. 23, 1891.

A band of fifty or more armed men, rustlers or not, depending wholly on one's point of view, had possession of the town of Buffalo. All the county officers were sympathetic. Independent roundup districts were laid out, and the small cattlemen and rustlers prepared to invade the ranges.[59] A petition by "citizens and taxpayers" was prepared, which demanded that the money held by the Live Stock Board and belonging to cattlemen suspected of being "rustlers" be remitted to the owners. This was sent to Cheyenne, but was disregarded by the Board on the ground that it did not express the real sentiments of the people; for honest folk had been forced to sign under threats of violence.[60]

Beset on all sides by a popular movement which threatened to destroy them, the members of the Association determined upon a course of action as drastic as it was unwise. As has been noted in another connection, the frontier has over and over again sanctioned the use of illegal and ruthless methods to preserve order and to protect life and property. The mistake of the Wyoming cattle companies, in the spring of 1892, in using such methods was in supposing that they were frontiersmen. These large cattle outfits, backed by outside capital, had lost all the characteristics of frontier enterprises; the small ranchmen and the granger represented what was left of the frontier on the northern ranges. Any illegal action against them, no matter how justifiable on the grounds that in no other manner could cattle stealing be suppressed, was sure to arouse the whole community.

On April 4, 1892, the Wyoming Stock Growers' Association held their regular annual meeting in Cheyenne. George Baxter, vice-president of the Association, presided in the absence of John Clay of Chicago, the president. Forty-three members answered the roll call and of these, all but eight represented cattle companies.

[59] *Cheyenne Daily Leader*, April 6, 1892.   [60] *Ibid.*, Feb. 24, 1892.

The business was mostly of a routine nature, but, among other things, the Association endorsed the action of the Live Stock Board in regard to the stock of suspected rustlers.[61] That the Board expected the Association to do more than merely approve of their action, is evident from the tenor of an address issued by the President of the Board, J. H. Hammond, a few days before the meeting. He described how the Wyoming cattlemen had come to that Indian-infested region and had risked their lives to build up a state. To his mind, they overcame obstacles far greater than those of the Pilgrim Fathers, for the Indians of Massachusetts had no kindly disposed Indian Bureau to supply them with the most up-to-date firearms. Now, the hardy pioneer, freed from the Indian danger, was about to be overwhelmed by the cattle thief. The address ended by pointing out what the frontier folk of Ohio, Illinois, and Indiana had done in days gone by to suppress cattle stealing in those regions.[62]

Whether the Association as a body determined upon the course of action subsequently followed, or whether the matter was decided upon unofficially, cannot be determined, as the records are not available. The Association never denied its connection with the events which occurred, nor was there any doubt in the popular mind as to the part played by that organization.[63]

The Association adjourned after a one-day session. On the following afternoon, there arrived from Denver a special train, which was parked in the Cheyenne yards. This train consisted of a chair car with curtains tightly drawn, a baggage car, a caboose, three stock cars loaded with saddle horses, and a flat car with wagons and camp paraphernalia on board. Citizens noted that no one was permitted to go

---

[61] *Cheyenne Daily Leader*, April 5, 1892.

[62] Minute Book of the Wyoming Board of Live Stock Commissioners, p. 9.

[63] The Executive Minutes of the Association contain no entries for 1892 save a statement that no meetings were held in that year (p. 93).

near this rather mysterious outfit. But everything had been so carefully attended to and so well planned, that when it pulled out over the Union Pacific tracks that night for the north, few in Cheyenne knew of its arrival or departure. Early the next morning, it arrived in Caspar, the point nearest to Johnson County on the Chicago North Western Railway. That day, the telegraph wires north of Caspar were cut, and southern Wyoming was left to guess what was going on in Johnson County.

The story of the "invasion" or "Johnson County War" is briefly told. In the closely shuttered car from Denver, were some twenty-five Texas hands, recruited at so much a head from the loafers around Denver. Just before the train left they were joined by a number of the leading members of the Association, lately adjourned. A "war" correspondent of the *Chicago Herald* was included in the party, apparently with the idea that the world at large must not lack information as to the conduct of the pioneers. After detraining, the party mounted their horses and set out for Johnson County, with Buffalo, the hotbed of rustlers and "anti-stock" men, as their objective.

On April 9, they succeeded in surrounding two alleged rustlers in an isolated ranch house and after killing one, as he went out to get water, they besieged the other through the day. Toward evening, they succeeded in setting fire to the cabin, and when the trapped man attempted to make a run for it, they riddled him with bullets. They then continued northward and, at nightfall, camped at a deserted ranch, fourteen miles from Buffalo.

The next morning, April 10, the frontier band started for town. They had not gone very far, when they were met by practically the whole population of Buffalo and the neighborhood. The news of the killings had got abroad and the whole countryside had turned out to repel the invaders.

There was a precipitate retreat to the ranch occupied the the night before. Here the cattlemen decided to make their stand against a formidable crowd of nearly two hundred: ranchmen, clerks, gamblers, rustlers, and what one correspondent euphemistically called, "men about town." It soon became clear that unless help arrived soon, the "invaders" would be beyond the help of the surgeon who had, with great foresight, been brought along.

News of the plight of some of Wyoming's leading citizens filtered into Cheyenne. Acting Governor Barber wired frantically to Washington, urging President Harrison to act immediately by ordering out the Federal troops stationed at Fort McKinney, twenty-five miles from Buffalo. Barber declared that a revolt was in progress and law and order must be restored. To the initiated, this meant that the leading citizens must be saved.

Accordingly, on April 13, three companies of Federal cavalry trotted upon the battlefield and the "war" was over. They arrived none too soon, for a movable platform, loaded with dynamite, was about to be pushed against the cabin where the cattlemen were besieged. The beleaguered, with great relief, surrendered to the officer in command who, it was suggested by observers, regretted that his command had arrived quite as soon as it did. Forty-six surrendered, twenty-five of whom were the hired Texans; the remainder were men prominent in the Association and the government of the state.[64]

Governor Barber insisted that the prisoners be conducted under guard of Federal troops to Cheyenne. This was prob-

[64] Among the names of those captured were the following:
  Major Walcott, commanding the expedition.
  W. C. Irvine, later president of the Association and member of the Board of Live Stock Commissioners.
  H. W. Davis, former president of the Association, and member of the legislature.

ably a wise move, for had they been turned over to the Johnson County authorities, they might have met with the same summary justice which they had meted out to others. Accordingly the leading citizens and the Texas gunmen arrived under Federal guard and were interned at Fort D. A. Russell awaiting further developments.[65]

In Johnson County the affair had so inflamed public sentiment that there was real danger that every head of stock belonging to the large companies would be slaughtered. The cattlemen were panicky. There was talk of martial law. President Harrison was induced to issue a proclamation ordering all unlawful assemblages to disperse before August 3, 1892. To those who knew what had occurred, the preamble to the proclamation made strange reading:

Whereas, by reason of unlawful obstructions, combinations and assemblages of persons, it has become impracticable, in my judgment, to enforce by the ordinary course of judicial proceedings the laws of the United States within the State and district of Wyoming, the United States marshal, after repeated efforts, being unable by his ordinary deputies or by any civil posse which he is able to obtain to execute the process of the U. S. courts; Now therefore, etc., etc.[66]

Good sense prevailed, however, and no untoward action

> Frank Canton, deputy United States marshal.
> W. J. Clarke, state water commissioner.
> Fred Hess, manager of the Bay State Cattle Co.
> H. E. Teschemacher, long a member of the executive committee of the Association and a Harvard graduate.

F. de Biller	Will Guthrie
E. W. Whitcomb	Chas. Ford
A. R. Powers	Ben Morrison
L. H. Parker	Scott Davis

[65] The above is taken from the newspaper account in the *Cheyenne Daily Leader*, which had become violently "anti-stock." Another account, equally partisan, is given by A. S. Mercer, *The Banditti of the Plains* (Cheyenne, 1894). The records of the Association contain no documentary material that might give the cattleman's side of this episode.

[66] Richardson, *Messages and Papers*, XX, 290.

was taken. The prisoners were finally turned over to the Johnson County authorities with the proviso that they should not be taken back to Johnson County for trial. The trial never took place. Johnson County was too poor to pay the cost of the trial, and it was extremely doubtful if an unprejudiced jury could be obtained in any part of the State. The material witnesses, two men who had spent the night in the cabin with the two murdered men and who were seized by the invaders before the shooting began, were taken in custody by the United States marshal on the charge of selling whiskey to the Indians and were taken to Omaha, where they disappeared. On August 10, 1892, the district court in Cheyenne decided that since Johnson County was unable to pay the expenses of a trial and since the court could not issue an order to force it to do so, there was nothing left but to admit the stockmen to bail on their own recognizances. This was done and the case ended there. The Texas cowboys were shipped off to Fort Worth and the ringleaders found reasons for leaving the state, some for good and others temporarily.[67]

The "invasion" made the ranges in the northeastern part of the State absolutely unsafe for the cattle of any of the persons implicated in the affair, and there was a general clearing out. A letter written to the secretary of the Association in May, 1893, from the manager of one of the large companies and one of the ringleaders, gives some idea of what took place.

As you know, we shipped everything last year [1892] we could find in all our brands, and the number of cattle we have left on the range is hard to calculate, for myself, I do not believe we have twenty-five head left. . . . . The Big Horn cattle belong to Windsor, Kemp and Co; but they may possibly be shipped by Mr. Robinson of the Bay State and shipped in his name as that was the only way we

[67] *Cheyenne Daily Leader*, Aug 11, 1892.

could get him to gather them; that was done to prevent the thieves in the Basin from making any kick about the Bay State gathering cattle that I have anything to do with. Windsor, Kemp and Co. cannot have more than 200 head in the old Power River Co. brands and the Big Horn Co. both together. . . . ." [68]

The political repercussions of the "war" were considerable. The Territory, hardly more than a satrapy of the Union Pacific during its earlier years, had been normally Republican. The close alliance of the Government with the Stock Growers' Association and the eagerness of the acting governor to restore order in Johnson County after the plans of the invaders miscarried gave their opponents the opportunity of linking up the so-called "Cheyenne cattle ring" with the Republican party organization. At the convention of the Democrats in the summer of 1892, those who were sympathetic with the cattlemen were read out of the party, the chairman, formerly an attorney for the Association, was deposed and a fusion with the Populist party consummated. The old Republican organization was placed on the defensive and spent most of its effort in denying that it had been in any way responsible for the events of the preceding spring. Senator Warren declared that he had no knowledge of the plan to invade Johnson County; ". . . . that he knew no more about it and contributed no more to it than a child unborn." [69]

Be that as it may, the old Republican machine narrowly escaped complete disaster. Although the new state cast its first electoral vote for Harrison, Weaver, the Populist candidate, missed carrying the state by only seven hundred votes. A fusion Democratic-Populist state ticket was elected

[68] Letter, Fred Hess to H. J. Ijams, May 31, 1893. Correspondence of Wyoming Stock Growers' Association, File, 1893

[69] *Cheyenne Daily Leader*, Nov. 1, 1892.

and a Silver Democrat was sent to Congress. The state legislature, composed of twenty-five Republicans, nineteen Democrats, and five Populists, refused to re-elect Warren for the Senate. After interminable wrangling, it adjourned without making a choice. The governor then appointed a Democrat as senator, but this gentleman resigned, and for two years Wyoming was represented by one senator only, Joseph M. Carey. Two years later, after the anti-stock excitement had died down, the Republican party returned to power.

In Montana, the decline of the range-cattle industry and the retreat of the range cattleman from the position which he occupied in the eighties were not accompanied by any such struggle. This was due, first to the fact that he had never occupied the preeminent position in Montana that he had in Wyoming — the mining and agricultural interests were, as we have noted, far too strong to allow such domination; and second, to the presence of newly opened Indian country in northern Montana where it was possible for the range system to continue for some time, undisturbed by the small ranchman and the farmer.

For some years to come, Miles City continued to be the cattle center of the Montana ranges. Not until 1902 was its position seriously challenged. At a meeting of the Montana Stock Growers' Association in that year, it was proposed that Helena be selected as the future meeting place instead of Miles City. It was declared that the sheepmen were taking over so much of the eastern Montana ranges that the old cow town was no longer the center of the cattle industry. The bulk of Montana cattle, it was argued, was being raised in the central and western sections of the state, in Jefferson, Madison, and Beaverhead counties. The retreat of the stock industry behind the defenses of privately owned hay lands and pastures in this area, where cattle had first been raised in Montana, had been accomplished. Only by a close vote, did the Association decide to continue its meetings in the eastern part of the state.[70]

Southern cattle continued to arrive on the northern Montana ranges for some years. As late as 1902, a Miles City paper noted that the movement of southern stock to the ranges had furnished more business than the railroads could handle. Old memories of other disasters were revived and the cattlemen of northeastern Montana were reminded of the ever present danger of overcrowding the range.[71] Two years later, the secretary of the Board of Live Stock Commissioners reported that this effort to continue the range-cattle system had been a signal failure. The remaining range had been overstocked and there had been heavy winter losses. A subsequent decline in the market had added to the ruin of the few remaining range outfits. Even though large areas of the range remained unoccupied, cattle raisers must, the report declared, get on a feeding and pasturage basis.[72]

Like the stock-growing industry, the expansion of the agricultural area of Montana was from the west, eastward. In the valley of the upper Yellowstone and its tributaries considerable farming areas had developed before the close of the eighties. The Gallatin Valley had, at a very early period, been a favored agricultural area and Bozeman a strong agricultural center. To the north, on the Missouri, Great Falls was founded in 1885. This town developed strong agricultural interests from the very start, its inhabitants looking with suspicion on the range-cattle industry in the neighborhood. One year after its founding, it was advertising the vast agricultural resources of northern Montana, declaring that any of the bench lands round about might be

---

[70] *Stock Growers' Journal* (Miles City), Apr. 17, 1902.

[71] *Northwest Live Stock Journal*, July 12, 1902.

[72] *Annual Report of the Secretary of the Board of Live Stock Commissioners*, 1904 (Helena, 1904), 7.

sown to wheat in the fall and a good crop produced without
irrigation. The crop of each successful farmer was hailed
as a proof that immense wheat fields would soon take the
place of the herds of the greedy "cattle kings." [73]

In 1893, the line of the Great Northern was completed
across the northern counties of the state. Settlement along
the line, particularly in the northeastern section, was slow.
Four years after its completion, 1897, an observer found
only herds of cattle in this, "the only exclusive range region
now left in this State." [74] Local papers hastened to inform
the visitor that, "in the older portions of our State, stock
are cared for as well as in Minnesota or other Mississippi
Valley States." [75]

Five years later, agricultural development along "the
High Line" was still in the future. Speaking before a crowd
in Great Falls in 1902, James J. Hill prophesied that
"churches and schools will be erected where now bands of
cattle and sheep roam; not that cattle and sheep are not all
right but farms are better."

"We want more railroads," some one in the crowd
shouted.

"You won't get them," Hill replied. "This railroad building is not fun. Develop your agricultural resources." [76]

Over on the Little Missouri, where once the great herds
of De Mores, Wibaux, and others ranged, reports came in
in the same year, 1902, that the settlers' dogs were making
it impossible for the old range outfits to operate and that
they were closing out. The range, thus deserted, was allowed
to go unoccupied, for the cattle of the incoming settlers
could not possibly fully utilize it for summer pasture. [77]

---

[73] *Great Falls Tribune*, Apr. 6, 1886.
[74] Letter by Prof. Thomas Shaw of the Minnesota Agricultural Experiment
Station, quoted in the *Rocky Mountain Husbandman*, Feb. 11, 1897.
[75] *Ibid.*   [76] *Stock Grower's Journal*, Aug. 21, 1902.
[77] *Stock Growers' Journal*, June 12, 1902.

A decade later, the Montana range still remained. The Secretary of the Live Stock Board in his report of 1913 declared that the open range still existed but that

. . . . the big stockman has been compelled to withdraw and the homesteader has not yet found himself in a financial condition to carry enough livestock to counteract the withdrawal of the large herds. . . . . Today, we could feed much more livestock on the ranches and farms during the winter than could be successfully run on the open range during the summer months.[78]

Thus the range cattleman disappeared.[79] Significantly enough, he lingered in those regions of northeastern Montana where the harried remnants of the great northern buffalo herd were last seen. Isolated ranches, where small cattlemen maintain themselves by combining winter feeding and a partial utilization of the range for summer pasturage, dry land farms, not a few deserted, and scattered settlements where irrigation has made agriculture safe and permanent, these have succeeded the great range outfits of other days. The story of the struggles of these groups, their partial successes and heartbreaking failures to utilize what the cattleman left, has yet to be told.

[78] *Annual Report of the Secretary of the Montana Board of Live Stock Commissioners*, 1913, p. 4.

[79] The dates of the last roundups on the northern ranges were about 1905-1906. An order issued by the Montana Board of Live Stock Commissioners in May, 1906, stated that, "Each roundup district shall constitute a local dipping district." (*Rocky Mountain Husbandman*, May 24, 1906). In Wyoming the Minute Book of the Board of Live Stock Commissioners, contains the following entry (p. 47) under the date of April 4, 1905:

"A call was made and published giving notice to the stockmen of a meeting to be held in the office of the Commission on April 4, 1905, to arrange the roundups and appoint commissioners therefor. As no stockmen appeared, the annual circular was not ordered.

"There being no further business, the meeting adjourned.

                                        Thomas F. Durbin, Sec'y."

BIBLIOGRAPHY

SOURCES

I. Manuscripts

The largest and most important collection of manuscript material used by the writer is to be found in the office of the Wyoming Stock Growers' Association at Cheyenne, Wyoming. This consists of the Minute Books of the Laramie County Stock Growers' Association and of the later Wyoming Association, the Minutes of the Executive Committee of the Wyoming Association, and the Correspondence of the Secretary of the Association. This latter is of unique value to the student of the range-cattle industry, for in one way or another, every phase of life on the High Plains during the range-cattle era is touched upon in the very extensive correspondence of the secretary of the largest of the range cattlemen's organizations.

After 1889 some of the functions of the Association were performed by the State Board of Live Stock Commissioners. The Minute Book, 1889-1906, is found in the office of the Board.

In the office of the Wyoming State Historian, the author had access to the Coutant Notes, a collection of notes on Wyoming history gathered by C. G. Coutant for his uncompleted *History of Wyoming*. Several fugitive manuscripts relating to Wyoming local history were used.

In Montana the Minute Book of the Stock Growers' Association is in the possession of the Montana State Historical Society. This book contains a full record of the meetings from 1885 to 1889.

II. Federal Documents

Annual Reports of the Secretary of Agriculture, 1870-1900; the Bureau of Animal Industry, 1884-1890; the Commissioner of the General Land Office, 1864-1890; the Commissioner of Indian Affairs, 1864-1888; the Secretary of the Interior, 1870-1890; the Secretary of War, 1859-1882.

Census, Eighth to the Fourteenth, 1860-1920.

Congressional Record.

Farmers' Bulletin No. 72, Department of Agriculture, "Cattle Ranges in the Southwest — A History of Exhaustion and Suggestion for Restoration," by H. L. Bently, 1898.

Forestry Service Report No. 72, "Live Stock Production in Eleven Far Western Range States," by W. C. Barnes and J. T. Jardine, 1916.

House and Senate Journals, 1870, 1874, 1884.

Indian Laws and Treaties (Washington, 1904), 2 vols.

Messages and Papers of the Presidents, compiled by James D. Richardson (Washington, 1899), 10 vols.

Opinions of the Attorney General, 1887.

The Public Domain by Thomas Donaldson (Washington, 1883).

Regulations of the Indian Bureau (Washington, 1884).

Synopsis of the Decisions of the Secretary of the Treasury, 1878-1879.

United States Statutes at Large.

Serial
Number

1189   Report of an expedition under Captain James L. Fiske to the Rocky Mountains, 1863, *House Ex. Doc.* No. 45, 38 Cong., Sess. 1.

1288   Report to the Secretary of War on the protection of the routes across the continent to the Pacific from molestation by hostile Indians, 1867, *House Ex. Doc.* No. 23, 39 Cong., Sess. 2.

1409   Report of the Senate Committee on Indian Affairs on petitions remonstrating against the imposition of a tax by the Cherokee Nation on cattle driven through their territory from Texas to the northern markets, 1870, *Sen. Report* No. 225, 41 Cong., Sess. 2.

1588} Report of the Select Senate Committee on transportation to the sea-
1589} board (Windom Report, 1874), *Sen. Report* No. 307, Parts 1 and 2, 43 Cong., Sess. 1.

1615   Report to the Secretary of War upon a reconnaissance of northwestern Wyoming made in the summer of 1873 by Captain William A. Jones, *House Ex. Doc.* No. 285, 43 Cong., Sess. 1.

1629   Report of Major General George A. Custer to the Secretary of War on an expedition to the Black Hills, 1874, *Sen. Ex. Doc.* No. 32, 43 Cong., Sess. 2.

1644   Letter of the Secretary of War concerning a military wagon road in Wyoming and Montana territories, 1874, *House Ex. Doc.* No. 22, 43 Cong., Sess. 2.

1818   Testimony taken before the House Committee on Public Lands on the surveys of the public lands, 1878, *House Mis. Doc.* No. 55, 45 Cong., Sess. 2.

1869   Information from the Commissioner of Indian Affairs to accompany the Presidential proclamation of April 26, 1879, against intrusion and unlawful occupation of the Indian Territory by white settlers, 1879, *Sen. Ex. Doc.* No. 20, 46 Cong., Sess. 1.

1923   Preliminary Report of the Public Land Commission, 1879-1880, *House Ex. Doc.* No. 46, 46 Cong., Sess. 2.

1975} Final Report of the Public Land Commission, 1879-1880, *House Ex.*
1976} *Doc.* No. 47, Parts 1-4, 46 Cong., Sess. 3.

1990   Report of the Treasury Cattle Commission on the lung plague of cattle or contagious pleuro-pneumonia, 1882, *Sen. Ex. Doc.* No 106, 47 Cong., Sess. 1.

Serial
Number

2076 Communication of the Secretary of the Interior to Congress with
      accompanying papers urging the necessity of stringent measures for
      the repression of the evasions and violations of the laws relating to
      the public lands, 1883, *Sen. Ex. Doc.* No. 61, 47 Cong., Sess. 2.

2159 Report of the House Committee on Public Lands on the repeal of the
      laws allowing preemption of the public land and amending the Home-
      stead Law, 1882, *House Report* No. 1834, 47 Cong., Sess. 2.

2165 Documents and correspondence relating to the leases of Indian lands
      to citizens of the United States for cattle grazing and other purposes,
      1884, *Sen. Ex. Doc.* No. 54, 48 Cong., Sess. 1.

2167 Report of the Commissioner of Indian Affairs relative to the leasing
      of lands upon the Crow Reservation in Montana, 1884, *Sen. Ex. Doc.*
      No. 139, 48 Cong., Sess. 1.

———   Report of the Commissioner of the General Land Office on the un-
      authorized fencing of the public lands, 1884, *Sen Ex. Doc.* No. 127,
      48 Cong., Sess. 1.

2200 Supplemental Report of the Treasury Cattle Commission on the in-
      spection of export cattle, 1883, *House Ex. Doc.* No. 44, 48 Cong.,
      Sess. 1.

2207 Report of the Treasury Cattle Commission on the disease of cattle
      in Kansas and Illinois, 1884, *House Ex. Doc.* No. 159, 48 Cong.,
      Sess. 1.

2257 Report of the House Committee on Public Lands on the unlawful
      occupancy of the public lands, 1884, *House Report* No. 1325, 48 Cong.,
      Sess. 1.

2261 Report of the Commissioner of Indian Affairs on the condition of the
      Cheyenne and Arapaho Indians, 1885, *Sen. Ex. Doc.* No. 16, 48 Cong.,
      Sess. 2.

2336 Annual Reports of the Government Directors of the Union Pacific
      Railroad to the Secretary of the Interior, 1864-1884, *Sen. Ex. Doc.*
      No. 69, 49 Cong., Sess. 1.

2341 Report of the Commissioner of the General Land Office relative to
      entries of the public land canceled for fraud, 1886, *Sen. Ex. Doc.*
      No. 225, 49 Cong., Sess. 1.

2356⎫ Report of the Select Senate Committee on the regulation of the trans-
2357⎭ portation of freights and passengers between the several states by
      railroad and water routes (Cullom Report, 1886), *Sen. Report* No. 46,
      Parts 1 and 2, 49 Cong., Sess. 1.

2362 Report of the Senate Committee on Indian Affairs on the condition
      of the Indians in the Indian Territory and other reservations, etc.,
      1886, *Sen. Report* No. 1278, 49 Cong., Sess. 1.

2445 Report of the House Committee on Public Lands on the ownership

Serial
Number

of real estate in the territories by foreigners, 1886, *House Report* No. 3455, 49 Cong., Sess. 1.

2560 Report of the Acting Commissioner of the General Land Office on the use of the public lands by cattle grazers, 1888, *House Ex. Doc.* No. 232, 50 Cong., Sess. 1.

2705 Report of the Senate Special Committee on the transportation and sale of meat products in the United States (Vest Report, 1888.), *Sen. Report* No. 829, 51 Cong., Sess. 1.

3021 Report on Agriculture by Irrigation in the western part of the United States at the Eleventh Census, 1890, by F. H. Newell, *House Mis. Doc.* No. 340, Part 20; 52 Cong., Sess. 1.

4766 Report of the Public Land Commission, 1905, *Sen. Doc.* No. 189, 58 Cong., Sess. 3.

III. STATE DOCUMENTS

Annual Reports of Territorial and State Auditors: Colorado, 1874-1889; Dakota, 1883-1889; Kansas, 1867, 1884; Montana, 1866-1888; Nebraska (Biennial) 1873-1890; Wyoming, 1869-1891.

Annual Reports of the Territorial Governors of Montana and Wyoming to the Secretary of the Interior, 1877-1888.

Annual Reports of the Secretary of the Montana Board of Live Stock Commissioners, the Veterinary Surgeon, and the Recorder of Marks and Brands, 1885-1890.

House Journal of the Montana Territorial Legislature, 1883.

Session Laws of the following territories and states: Colorado, Dakota, Kansas, Missouri, Montana, Nebraska, Texas, and Wyoming.

Laws of the Cherokee Nation (St. Louis, 1875).

Transactions of the Nebraska State Board of Agriculture, 1873-1880.

Wyoming Agricultural Statistics: Joint Bulletin issued by the United States Department of Agriculture and the Wyoming State Department of Agriculture (Cheyenne, 1925).

IV. CASES

Canfield et al. *v.* United States, 66 *Federal* 101.

Railroad *v.* Husen, 95 *United States* 465.

Stewart *v.* Hunter, 8 *American State Reports* 270.

United States *v.* Douglas, Willan-Sartoris, 3 *Wyoming Reports* 288.

United States *v.* Loving, 2 *Sawyer* 148.

United States *v.* Mattocks, 34 *Federal* 715.

V. GREAT BRITAIN

The Law Reports, the Public General Statutes, 47-48. Victoria, 1884.

Parliamentary Papers. House of Commons, 1880, Vol. 18, Cd. No. 2678. Report of the Agricultural Interests Commission.

House of Commons, 1878-1879, Vol. 58, No. 111. Correspondence connected with the Detection of Pleuro-Pneumonia among Cattle landed in Great Britain from the United States.

House of Commons, 1884-1885, Vol. 73, Cd. No. 4372. Contagious Diseases (Animals) Act 1878 (Ireland).

House of Commons, 1884-1885, Vol. 73, No. 19, Sess. 2. Contagious Diseases (Animals) Act 1884 (Canada).

VI. NEWSPAPERS

American Agriculturist, 1865, 1868, 1871.
Bozeman Avant Courier (Bozeman, Montana), 1874-1885.
Bozeman Chronicle (Bozeman, Montana), 1880-1885.
Breeder's Gazette (Chicago, Ill.), 1881-1888; 1912.
Butte Miner (Butte, Montana), 1879-1882.
Cheyenne Daily Leader (Cheyenne, Wyoming), 1867-1892.
Cheyenne Daily Sun (Cheyenne, Wyoming), 1880-1888; 1891-1892.
Chicago Tribune (Chicago, Ill.), 1884.
Glendive Times (Glendive, Montana), 1882-1885.
Great Falls Tribune (Great Falls, Montana), 1885-1895.
Helena Daily Herald (Helena, Montana), 1874-1882.
Helena Daily Independent (Helena, Montana), 1879-1884.
London Times (London, Eng.), 1884.
London Economist (London, Eng.), 1883-1884; 1886.
Miles City Weekly Press (Miles City, Montana), 1882-1884.
Missoulian (Missoula, Montana), 1874-1875.
Montanian (Virginia City, Montana), 1871-1872.
National Live Stock Journal (Chicago, Ill.), 1871-1874.
New Northwest (Deer Lodge, Montana), 1874.
New York Times (New York), 1884.
Northwestern Live Stock Journal (Cheyenne, Wyoming), 1886-1887.
River Press (Fort Benton, Montana), 1882-1884.
Rocky Mountain Husbandman (Diamond City, White Sulphur, Great Falls, Montana), 1875-1890.
Stock Growers' Journal (Miles City, Montana), 1902-1903.

VII. REPORTS OF BOARDS OF TRADE, CORPORATION REPORTS AND OTHER CONTEMPORARY MATERIAL

Atchison, Topeka & Santa Fé Railroad, Annual Reports of the Directors.
The Big Laramie Land, Cattle and Improvement Company of Wyoming Territory (No date, *circa* 1886).
Chicago, Annual Reports of the Chicago Board of Trade (1865-1890).

Central Pacific Railroad, *Nelson's Pictorial Guide-book* (New York, *circa* 1875).

Denver, Annual Reports of the Denver Board of Trade, 1872-1880.

Missouri, Kansas and Texas Railroad, *Reports and Statements*, 1879 (St. Louis, 1879).

Montana Stock Growers' Association, *Brand Book*, 1885.

National Association of Stock Breeders, *Proceedings of the First Annual Convention*, 1883 (Chicago, 1883).

National Cattle and Horse Growers' Association of the United States, *Proceedings of the First Annual Meeting*, 1884 (St. Louis, 1884).

National Association of Cattle Growers, *Proceedings of the Third Annual Convention*, 1885 (Chicago, 1885).

National Consolidated Cattle Growers' Association of the United States, *Proceedings of the National Convention at Kansas City*, 1887 (Chicago, 1887).

New York, Annual Reports of the New York Produce Exchange. 1879-1890.

Omaha, Annual Reports of the Omaha Board of Trade, 1870-1890.

Union Stock Yards, *Annual Review of the Live Stock Trade as transacted at the Union Stock Yards*, by E. Griffiths (Chicago, 1872).

Annual Reports of the Directors of the Union Stock Yard and Transit Company. 1870-1890.

Wyoming Stock Growers' Association, *Brand Books*, 1881-1886, in the possession of the Wyoming State Historian.

Wyoming Stock Growers' Association, *By-Laws and Reports*, 1884 (Cheyenne, 1884).

VIII. Journals and Other Contemporary Accounts

ALDRIDGE, REGINALD, *Life on a Ranch; Ranch Notes in Kansas, Colorado, the Indian Territory and Northern Texas* (London, 1884).

ALLEN, LEWIS F., *American Cattle, Their History, Breeding, and Management* (New York, 1881).

ARMOUR, PHILLIP D., *The Present Condition of the Live Cattle and Beef Markets in the United States* (Chicago, Ill., 1889).

AUDUBON, JOHN W., *Western Journal, 1849-1850* (Cleveland, 1906).

BEADLE, JOHN H., *The Undeveloped West, or Five Years in the Territories* (Philadelphia, 1873).

BOWLES, SAMUEL, *Our New West* (Hartford, 1869).

BRISBIN, JAMES S., *Beef Bonanza, or How to Get Rich on the Plains* (Philadelphia, 1881).

CROSS, FRED J., *The Free Lands of Dakota* (Yankton, 1876).

DILKE, CHARLES W., *Greater Britain*, 2 vols. (London, 1869).

FARNHAM, T. J., "Travels in the Great Western Prairies, 1839" in *Early Western Travels*, Thwaites, R. G., ed. (Cleveland, 1906), Vol. XXVIII.

GREELEY, HORACE, *An Overland Journey from New York to San Francisco in the Summer of 1859* (New York, 1860).

JULIAN, GEO. W., "Our Land Policy," *Atlantic Monthly*, Vol. XLIII, pp. 325-337, March, 1879.

LATHAM, HENRY, *Trans-Missouri Stock Raising: the Pasture Lands of North America* (Omaha, 1871).

McCOY, JOSEPH G., *Historic Sketches of the Cattle Trade of the West and Southwest* (Kansas City, Mo., 1874).

MACDONALD, JAMES, *Food from the Far West* (London, 1878).

MERCER, ASA S., *The Banditti of the Plains* (Cheyenne, 1894).

NIMMO, JOSEPH, *The Range and Ranch Cattle Business of the United States* (Washington, 1885).

——— "The American Cowboy," *Harpers Monthly*, LXXIII, Nov. 1886.

OWEN, MAJ. JOHN, *The Journals and Letters of Major John Owen, Pioneer of the Northwest, 1850-1871*, Paul C. Phillips, ed. (New York, 1927).

PALLISER, JOHN, *Solitary Rambles and Adventures of a Hunter in the Prairies* (London, 1853).

PALMER, JOEL, "Journal of Travels over the Rocky Mountains, 1845-1846," in *Early Western Travels*, Thwaites, R. G., ed. (Cleveland, 1906), Vol. XXX.

PORTER, ROBERT P., *The West from the Census of 1880* (Chicago, 1882).

POWELL, JOHN W., *Report on the Arid Regions of the United States* (Washington, 1879).

RUSLING, JAMES F., *The Railroads! The Stockyards! The Eveners! An Exposé of the Railroad Ring* (Washington, 1878).

STRAYHORN, R. E., *Handbook of Wyoming and Guide to the Black Hills and Big Horn Regions* (Cheyenne, 1877).

——— *The Resources of Montana* (Helena, 1879).

STUART, GRANVILLE, *Montana as it is, 1865* (New York, 1865).

TRIGGS, J. H., *History of Cheyenne and Northern Wyoming* (Omaha, 1876).

TOPPING, E. S., *The Chronicles of the Yellowstone* (St. Paul, 1883).

WALKER, FRANCIS A., *The Indian Question* (Boston, 1874).

SECONDARY WORKS

ABEL, ANNIE H., *The American Indian under Reconstruction* (Cleveland, 1925).

ADAMS, ANDY, *Log of a Cowboy* (Boston, 1903).

ADAMS, R., "Public Range Lands — A New Policy Needed," *American Journal of Sociology*, Vol. XXII, p. 324, Nov. 1916.

ARMOUR'S Live Stock Bureau, "Cattle Trails in Live Stock Market Development," *The Monthly Letter to Animal Husbandmen* (Chicago, Ill.), Vol. VII, No. 1, April, 1926.

BARNES, W. C., *Winter Grazing Grounds and Forest Ranges* (Chicago, 1913).

BLACKMAR, FRANK W., *Spanish Institutions of the Southwest* (Baltimore, 1891).

BUCHANAN, J. R., "The Great Railroad Migration into Northern Nebraska," in the *Proceedings and Collections of the Nebraska Historical Society* (Lincoln, Neb., 1907), Vol. XV, pp. 25-34.

CLAY, JOHN, *My Life on the Range* (Chicago, 1924).

CLEMEN, RUDOLPH A., *The American Livestock and Meat Industry* (New York, 1923).

COMAN, KATHERINE, *Economic Beginnings of the Far West*, 2 vols. (New York, 1912).

CONNOR, L. G., "A Brief History of the Sheep Industry of the United States," *Annual Report of the American Historical Association*, 1918, Vol. I, pp. 93-197.

CONOVER, MILTON, *The General Land Office* (Baltimore, 1923).

Contributions of the Historical Society of Montana (Helena, Mont.), Vols. 1-9, 1876-1923.

COOK, JAMES H., *Fifty Years on the Old Frontier* (New Haven, 1923).

COUTANT, C. G., *History of Wyoming* (Laramie, 1899), Vol. I.

CRUISE, J. D., "Early Days of the Union Pacific," *Collections of the Kansas State Historical Society* (Topeka, 1910), Vol. XI, pp. 529-550.

DALE, EDWARD E., "The History of the Ranch Cattle Industry in Oklahoma," *Annual Report of the American Historical Association*, 1920, pp. 309-322.

DAVIS, JOHN P., *The Union Pacific Railway* (Chicago, 1894).

DUFFIELD, JOHN, "Driving Cattle from Texas to Iowa, 1866," *Annals of Iowa* (Iowa City), Vol. XIV, No. 4, pp. 242-262, April, 1924.

FARMER, HALLIE, "The Economic Background of Frontier Populism," *Mississippi Valley Historical Review*, Vol. X, No. 4, pp. 406-427, March, 1924.

GANNETT, HENRY, *The Boundaries of the United States* (Washington, 1904).

GARLAND, HAMLIN, *A Son of the Middle Border* (New York, 1917).

GITTINGER, ROY, *The Formation of the State of Oklahoma, 1803-1906* (Berkeley, 1917).

HAGEDORN, HERMAN, *Roosevelt in the Bad Lands* (Boston, 1921).

HANEY, LEWIS H., *A Congressional History of the Railways 1850-1887*, Bulletin No. 211 of the University of Wisconsin Economics and Political Science Series, Vol. III, No. 2 (Madison, 1908).

HARGER, CHARLES M., "Cattle Trails of the Prairies," *Scribner's*, Vol. XI, pp. 732-742, June, 1892.

HAZELTON, JOHN M., *History and Handbook of Hereford Cattle and Hereford Bull Index* (Kansas City, Mo., 1925).

HEBARD, G. R., and BRININSTOOL, E. A., *The Bozeman Trail* 2 vols. (Cleveland, 1924).

HIBBARD, BENJ. H., *History of the Public Land Policies* (New York, 1924).

HILL, ROBERT T., *Public Domain and Democracy*, Studies in History, Economics and Public Law, ed. by the Faculty of Political Science of Columbia University, Vol. XXXVIII, No. 1 (New York, 1910).

HOPPER, SILAS L., "Nebraska City to California, April-August, 1863," *Annals of Wyoming* (Cheyenne), Vol. III, No. 2, pp. 117-126, Oct., 1925.

HOUGH, EMERSON, *Passing of the Frontier* (New Haven, 1918).

HUNTER, J. M., ed., *The Trail Drivers of Texas* (2nd ed. Nashville, 1925).

JAMES, WILL, *Cowboys, North and South* (New York, 1924).

KENDRICK, J. B., "The Texas Trail," *Wyoming Historical Society Miscellanies* (Cheyenne, 1919), pp. 41-49.

KOHRS, CONRAD, "A Veteran's Experience in the Western Cattle Trade," *Breeder's Gazette* (Chicago, Ill.), Dec. 18, 1912, p. 1328.

KUHN, BERTHA M., "The W-Bar Ranch on the Missouri Slope," *Collections of the State Historical Society of North Dakota* (Grand Forks), Vol. V, pp. 153-166.

LANG, LINCOLN, *Ranching with Roosevelt* (Philadelphia, 1926).

LOVE, CLARA M., "Cattle Industry of the Southwest," *Southwestern Historical Quarterly* (Austin, Texas), Vol. XIX, No. 4, pp. 370-399, April, 1916.

MORRIS, R. C., "The Notion of the Great American Desert East of the Rockies," *Mississippi Valley Historical Review*, Vol. XIII, No. 2, pp. 190-200, Sept., 1926.

PALLADINO, L. B., *Indian and White in the Northwest* (Baltimore, 1894).

PAXSON, FREDERIC L., *History of the American Frontier* (New York, 1924).

———— *The Last American Frontier* (New York, 1910).

———— "The Cow Country," *American Historical Review*, Vol. XXII, pp. 65-82, Oct. 1916.

———— "The Pacific Railroads and the Disappearance of the Frontier in America," *Annual Report of the American Historical Association*, 1907, Vol. I, pp. 107-118.

PELZER, LOUIS, "A Cattlemen's Commonwealth on the Western Range," *Mississippi Valley Historical Review*, Vol. XIII, No. 1, pp. 30-49, June, 1926.

PLUNKETT, SIR HORACE, *The Rural Life Problem of the United States* (New York, 1911).

POOR, HENRY V., *Manual of the Railroads of the United States, 1868-1902*.

RIEGEL, ROBERT I., *Story of the Western Railroads* (New York, 1926).

RIPLEY, W. Z., *Railway Problems* (New York, 1913).

ROLLINS, P. A., *The Cowboy* (New York, 1922).

ROOSEVELT, THEODORE, *An Autobiography* (New York, 1921).

SANBORN, JOHN B., *Congressional Grants of Land in Aid of Railways*, Bulletin No. 30 of the University of Wisconsin. Economics and Political Science Series (Madison, 1899), Vol. II, No. 3.

SCHMIDT, L. B., "The Westward Movement of the Corn-Growing Industry in the United States," *Iowa Journal of History and Politics* (Iowa City), Vol. XXI, No. 1, pp. 112-141, Jan., 1923.

—— "The Westward Movement of the Wheat Growing Industry of the United States," *Iowa Journal of History and Politics*, Vol. XVIII, No. 3, pp. 396-412, July, 1920.

SEMPLE, ELLEN C., *American History and Its Geographic Conditions* (Boston, 1903).

SLOAN, W. K., "Autobiography," *Annals of Wyoming* (Cheyenne), Vol. IV, No. 1, pp. 235-264, July, 1926.

SMALLEY, E. W., *History of the Northern Pacific Railroad* (New York, 1883).

STONE, A. L., *Following Old Trails* (Missoula, Mont., 1913).

STEWART, GEORGE, "A Land Policy for the Public Domain," *Economic Geography* (Worcester, Mass.), Vol. I, No. 1, pp. 89-106, March, 1925.

STUART, GRANVILLE, *Forty Years on the Frontier*, Phillips, Paul C., ed., 2 vols. (Cleveland, 1925).

TRIMBLE, WILLIAM, *The Mining Advance into the Inland Empire*, Bulletin No. 638 of the University of Wisconsin History Series (Madison, Wis.), Vol. III, No. 2.

—— "Historical Aspects of the Surplus Food Production of the United States, 1862-1902," *Annual Report of the American Historical Association*, 1918, Vol. I, pp. 223-239.

TROTTMAN, N., *History of the Union Pacific* (New York, 1923).

TURNER, FREDERICK J., *The Frontier in American History* (New York, 1920).

VAN HISE, CHARLES H., *Conservation of the Natural Resources of the United States* (New York, 1926).

VINOGRADOFF, P. G., *The Growth of the Manor* (London, 1905).

VOLWEILER, A. T., "Roosevelt's Ranch Life in North Dakota," *Quarterly Journal of the University of North Dakota*, Vol. IX, No. 1, pp. 31-49, Oct., 1918.

WHITCOMB, E. W., "Reminiscences of a Pioneer," First *Biennial Report* of the State Historian of the State of Wyoming (Laramie, 1920).

WILSON, M. L., "Evolution of Montana Agriculture in its Early Period," *Proceedings of the Mississippi Valley Historical Association*, Vol. IX, pp. 429-440, 1915-1918.

WYOMING STOCK GROWERS' ASSOCIATION, *Letters from Old Friends and Members of the Wyoming Stock Growers' Association* (Cheyenne, 1923).

YOUNGBLOOD, B., "The Position of Ranching in our National Economy," *Southwestern Political Science Quarterly* (Austin, Texas), Vol. III, No. 1, pp. 16-24, June, 1922.

INDEX

Agricultural frontier, Missouri, 1; Oregon, 6; eastern Kansas and Nebraska, 19, 24-25, 37, 45; barrier to Texas cattle, 37, 163, 177; in western Montana, 56-57; beginnings in Wyoming, 68; reaches semi-arid region, 84; transition in western Montana, 87; supplies stock to the western range, 92; contrast with cattleman's frontier, 115; estray law, 127; extension in Wyoming, 189, 226, 240-245; land policy adapted to, 194; recoil in Kansas and Nebraska, 225; development after 1890, 227-237, 256-258. *See also* Feeders, Irrigation, Land policy, and Ranches

Alfalfa, forage crop on the range, 227

Ancenny, Charles, 122

Arkansas Cattle Company, extensive fencing by, 191

Ashley, W. H., 5

Assiniboine Indians, 63

Atchison, Topeka and Santa Fé Railroad, 90

Barber, Amos W., 251-252

Beaverhead Valley, Mont., refuge of early Wyoming cattlemen, 12, 14-15; early agricultural section, 56-57, 231

Big Horn Basin, Wyo., exposed to Sioux attack, 61; grazing area, 61; mining expedition to, 72-73; advertisement of, 76

Big Horn City, Mont., 77

Big Horn Mining Association, 72-73

Big Horn Mountains, western boundary of the Sioux hunting grounds, 61

Bismarck, Dakota Terr., terminus of the Northern Pacific, 65; Montana cattle driven to, 80-81

Black Hills, discovery of gold in, 75; effect of mining boom on the Indian problem, 76-77

Blackfeet Indians, extent of reservation, 63; reduction of reservation, 64; dangerous to cattlemen, 142, 144-147; rights to the Musselshell country, 145-146; face starvation, 146

Black-list, used by stock growers' associations, 149-150, 156-157, 246-247

"Blue Bucket Diggings," 71

"Book count," 101, 103

Boycott, used by cattlemen's organizations, 185-189

Bozeman, Mont., Yellowstone expedition organized at, 73-74; agricultural center, 256

Bozeman Trail, first Texas cattle driven into Montana over, 21; connected Montana with the Oregon Trail, 60; Indian war along, 60-61; forts on, 66-67; western demand to reopen, 68, 76-77; effort to reopen, 74

Brands, early regulations in Texas as to, 33; necessity for regulations concerning, 115; legislation as to, 125-127, 137; alteration of, 126, 135, 148-149; regulations as to calves on the range, 134-135; regulations of the Wyoming Stock Growers' Association concerning,

269

in Scotland and England, 98-101;
evils of, 101-105

Splenetic fever, see Texas fever

Stock detectives, in Texas, 118; employment of, by associations, 149-
151; enmity for associations increased by the use of, 157-158

Stock inspection, Wyoming Association's methods of, 151-155; a state
function in Montana, 161-162; Wyoming effort to check rustling
through, 246

Story, Nelson, brings first herd of
Texas cattle to Montana, 21

Stuart, Granville, on the expansion
of Montana cattleman's frontier
eastward, 88-89; on the control of
grazing, 182; on the range-cattle
business, 221

Sturgis, Thomas J., secretary of the
Wyoming Stock Growers' Association, on shipping Texas cattle by
rail, 165; member of legislative
committee of the National Convention of Stock Growers, 170; on the
future of the range-cattle industry,
206-207

Sun River, Mont., early cattleman's
frontier, 54, 57; organization of
cattlemen on, 123, 146-147

Survey of the public domain, ill
adapted to the arid West, 18;
Powell proposes to change system
of, 196; system as applied to railroad grants, 214

Sutherlin, E. H., editor of the Rocky
Mountain Husbandman, 122

Swan Land and Cattle Co., organization of, 97-98; illegal fencing by,
192; failure of, 222

Sweetgrass River, Mont., expansion
of cattleman's frontier eastward
to, 77

Sweetwater River, Wyo., on the route

of the Oregon Trail, 5; mines in
the vicinity of, 40-41, 71

Texas, conditions following the Civil
War in, 28-29; early cattle drives
from, 30-32; cattlemen's associations in, 32; passes laws regulating range-cattle industry, 33-34;
supplies cowboys for the northern
range, 48; railroads tap the cattle
ranges of, 90; costs of raising cattle in, 139 footnotes; urges a National Cattle Trail, 139-140, 179-
180; opposition to Animal Industry Bill in, 170-173; representation
in the National Cattle Growers'
Convention, 179

Texas cattle, arrival of first herd in
Montana, 21; origin of, 26;
description of, 26-27; early market for, 27-28; early prices of, 28;
sold to Iowa farmers, 31; arrive
in Kansas, 31-32, 36-37; northern
range as a market for, 38; arrival
in Colorado, 39-40; arrival in Wyoming, 42-43; early shipments to
Chicago, 44-45; arrival in Montana, 54; insufficient to meet the
demand of the northern ranges,
89; northern ranges receive rail
shipments of, 91; threaten the improved cattle of the northern ranges, 138-140

Texas drive, Abilene, a terminus of,
25, 31; character of, 26; routes
taken by, 30-31, 36-37; numbers in,
31-32; organization of, 32; regulations in the Indian Territory of,
34-36; Kansas farmer an obstacle
to, 37, 177-178; Ogallala, a terminus of, 51-52; railroads sup-
plant, 91; cooperative character
of, 118; effect of quarantine regulations on, 163-164

Texas fever, quarantine laws to pre-

tion in, 222-226; increase in forage crops in, 227-228; size and number of farms in, 234-236; granger invasion of, 240-244; "farm homes for poor men" in, 243-244; last stand of the "cattle kings" in, 248-255; last roundups in, 258 footnote

Wyoming Hereford Association, description of ranch of, 204

Wyoming Stock Graziers' Association, first cattleman's organization in Wyoming, 119

Wyoming Stock Growers' Association, organization of, 120-121; compared with the Montana Association, 125; jurisdiction extends outside of Wyoming, 132; power of, 135-137, 154-158; relations with the Indian Bureau, 142-144; blacklists of, 149-150; instructions to roundup foremen, 151; inspection system of, 152-153; frontier methods of, 157-158; Chicago commission men blacklisted by, 173;

represented in the National Cattle Growers' Convention, 179; opposition to the National Cattle Trail, 180; control ranges by limiting membership, 186-188; action on the Powell Report, 199-200; decline of, 238-239; attack of rustlers and grangers on, 240-245; effort to control newspapers, 247; part played in the Johnson County War, 247-255

Yellowstone Park, Wyo., road from the Union Pacific to, 63

Yellowstone River, Mont., natural outlet for Montana, 62; effort to open a route along, 73-74; advertised by military campaigns, 76; early settlement on, 77; Northern Pacific arrives on, 84; new ranges on, 87

Yellowstone Wagon Road and Prospecting Expedition, 73-74

PHOENIX BOOKS
in History